THE COMPLETE BOOK OF
LOCKS &
LOCKSMITHING
-2ND EDITION

BY C. A. ROPER

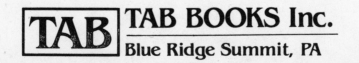
TAB BOOKS Inc.
Blue Ridge Summit, PA

SECOND EDITION

ELEVENTH PRINTING

Printed in the United States of America

Library of Congress Cataloging in Publication Data

Roper, C. A. (Carl A.)
The complete book of locks & locksmithing.

Rev. ed. of: The complete handbook of locks & lock-smithing. 1st ed. c1976.
Includes index.
1. Locksmithing.   I. TItle.   II. Title: Complete hand-book of locks and locksmithing.
TS520.R58 1983   683'.32   82-19386
ISBN 0-8306-0130-9
ISBN 0-8306-1530-X (pbk.)

TAB BOOKS Inc. offers software for sale. For information and a catalog, please contact TAB Software Department, Blue Ridge Summit, PA 17294-0850.

Questions regarding the content of this book should be addressed to:

Reader Inquiry Branch
TAB BOOKS Inc.
Blue Ridge Summit, PA 17294-0214

Cover illustration by Al Cozzi.

Contents

Acknowledgments

A book such as this could not have been written without the dedication and assistance of many individuals and companies within the locksmithing industry. While too numerous to mention all by name, I should like to single out a number of individuals and companies that went above and beyond the "call of the writer" in providing assistance and information: Bill O'Connor, Medeco Security Locks; Mary Sanowskis, Preso-Matic; Frank Sushan, Sentry Door Lock Guards; Ron Levin, Lock Corporation of America; Herbert Stein, President of Star Key & Lock Mfg. Co.; Bruce Williams, Vice President, The M. K. Morse Co.; Jack Buzard, Adams Rite Mfg. Co.; James Faherty, Dominion Locks; Gary Paley, Trine Consumer Products Division, Square "D" Co.; Ralph Sittnik, President, Security Engineering, Inc.; Robert Psolka, Taylor Lock Co.; and Thomas Nazziola, Simplex Security Systems, Inc. Each of these people's endeavors and participation in this book is gratefully appreciated. And to Liz Akers at TAB Books who provided assistance and answered innumerable questions concerning the book contents, a special "thank you."

Finally, for my wife, Lynda, and the kids, for surviving still another major writing effort, a very special "thank you" is also in order.

Introduction

This handbook on locksmithing is written for the student, the professional locksmith, and for everyone who is interested in the fascinating world of locks.

There is enough material here to get you started in the business of selling, servicing, and installing locks. Besides the challenge of the work, locksmithing is a very lucrative profession. According to the *U.S. Occupation Outlook Handbook* and the *Locksmith Ledger*, a self-employed locksmith can look forward to earning twice the hourly wage of an auto mechanic and three times as much as a watchmaker. Locksmithing is rewarding in other ways; few other crafts offer such prestige and trust.

The amateur, the student of lock technology, and the collector will find much that is rewarding here. One chapter traces the evolution of the lock from simple Egyptian bolts through Greek and Roman designs, locks of the Middle Ages and the Renaissance, the latchstrings of the American frontier, and concludes with the great 19-century revolution in lockmaking. Other chapters carry the story to modern times, including ultra-high security locks used in military installations.

I have tried to include everything that you need to know. Each of the various lock mechanisms—combination, warded, lever, disc, wafer, and pin-tumbler—is discussed in detail with an eye to troubleshooting and servicing. Chapters are devoted to interior and exterior locksets, with special emphasis on the most popular brands. While locksmithing requires surprisingly few tools, each tool that you will need is described and, in most cases, illustrated. Instructions are given on how to make your own tools and how to choose and buy factory-made tools.

Professionals will welcome the chapter on master key systems, the in-depth information on Schlage and Corbin locksets, and the discussion of Medeco and other specialty locks. Vending-machine, filing-cabinet, post-office box, and automobile locks are illustrated and discussed in detail.

Electrically operated locks and/or strike type units are becoming increasingly popular and in demand. Therefore, a new chapter in this second edition focuses on this subject. This tremendously expanding business section for the locksmith can include installation, maintenance and repair of various types of access and exit control and/or monitoring devices. Those of you who contemplate entering this phase of specialization and adding it to your business will find it a welcome added chapter. For locksmiths who have only an occasional need for such information, the chapter is at the rear of the book, so you can concentrate on the more basic area, in addition to the primary reference material and details necessary to your day-to-day business.

Chapter 1

A Short History of Locks and Keys

Locks and keys have been around for centuries. They have evolved with man as a natural part of his existence, signifying his continual need for protection of life and property. Every civilization in history understood the need for security, and their understanding spawned a persistent search for more reliable locks, and more sophisticated keys.

EGYPT

Long before the glories of Greece, Rome, and the Greco-Roman empire, the Egyptians developed a reliable lock that operated in essentially the same way our locks do today. They invented the pin-tumbler lock and key. Old paintings inside pyramids give us a general idea of what the lock was like (Fig. 1-1). The internal form of the lock was similar to that reinvented in the mid-1800s by Linius Yale, Sr., in the United States.

The lock body was made of wood and mounted securely to the door. The bolt passed through the lock body and into a bracket mounted on the wall. In the locked position, hardwood or iron pins dropped into holes in the bolt, fixing it securely in the lock body. The key was no more than a length of wood fitted with pins that matched those in the lock body. It was inserted through the hollow end of the bolt (the bolt hole, forerunner of the keyhole) and raised to disengage the pins from the bolt. The bolt was then free to slide. The size, pattern, number, and length of the key pins varied, even as they do today.

GREECE

Most Greek doors pivoted at the center and were secured by bolts from the inside. In the few cases where locks were used, they were very primitive. The key was a large crescent-shaped affair, resembling a modern sickle. To work the bolt, the owner inserted the key into the keyhole and gave it a twist. The tip of the key aligned with a hole in the bolt and moved it left or right, locking or unlocking the door.

ROME

Unfortunately, most Roman locks have been destroyed by time, but the few keys and locks that remain give testimony to the Roman genius. The basic lock was of the warded type. (A *ward* is a projecting ridge in a lock or on a key designed to permit only the correct key to be inserted in a lock.) Many locks are still made on this principle. In addition, the Romans made ingenious attempts to disguise the keyway as ornamentation or to hide it altogether. One famous Roman lock is in the shape of a fish. The keyway is revealed when one of the fins is turned.

Roman innovations included the use of wards in their own iron case, the metal key, the spring-loaded bolt, the spring-operated padlock, and the first true (i.e., removable) padlock. One curious aspect of Roman lock lore is that the keys were designed to be worn as rings because togas did not have pockets (Fig. 1-2).

After the fall of Rome, Europe was swept by Barbarian hordes moving south and east from the Rhineland and Scandinavia. Brute force was the order of the day, and locks meant little. Men put their faith in arms and fortresses.

This period passed as the Barbarians learned from their victims and settled down into the routines of farming and fishing. The Vikings who settled in Brittany evolved a new kind of civilization, with a king at the apex and power radiating downward from him into various levels of nobility. This social pattern, known as *feudalism*, gave the nobility access to wealth and property, and the art of locksmithing again flourished.

Fig. 1-2. Roman keys were worn as rings.

EUROPE

In Europe, keys were made that could move about a post and shift the position of a movable bar (the locking bolt). In its various forms it worked quite well. The first obstacles to unauthorized use of the lock were various internal wards. Medieval and Renaissance craftsmen made warded locks and improved upon them. Some were very complex, using many interlocking wards and complicated keys. Since many of these wards could be easily bypassed by almost anyone, newer methods had to be devised.

Later, chests were made entirely of metal and the locksmith reasoned that if one bolt was secure, then a chest with eight, ten, twelve, or even fifteen locking bolts would be more secure. Thus, craftsmen created elaborate internal mechanisms to allow the many bolts to be shifted by one or two separate keys. Levers, springs, ratchets, and pinions were employed to do this job. The locksmiths also installed separate locks for separate bolts; the locks had false keyholes and safety devices, such as spring-loaded knives that would injure or kill the thief. Hidden keyholes were extremely popular among the merchant and upper classes.

In France, the treatise *The Art of the Locksmith* (1767) was published, describing examples of the *tumbler* or *lever lock*. Exactly who invented it is unknown, but credit is long overdue. As locksmithing advanced, multiple-lever locks came into being. Two, three, six, and more levers were used. Each lever had to be lifted; when all were in proper alignment, the bolt could be moved, opening or closing the lock.

Fig. 1-1. An ancient Egyptian lock. Raising the key brought the pins to the shear line and released the bolt. The same principle was used some 400 years later on the Yale lock.

2

ENGLAND

The English also worked on newer and better locking devices. Incentive was given in the form of cash awards and honors to those who could successfully open these newer and more complex locks. In the forefront of lock designing were three Englishmen: Robert Barron, Joseph Bramah, and Jeremiah Chubb.

Robert Barron patented the double-action tumbler lock in 1788. Like others before it, this lock employed a series of lever tumblers pegged at one end (Fig. 1-3). One side of these tumblers engaged a notch in the bolt. The key bore against the opposing surface. Notches on the key corresponded to the individual tumblers. The width of the key-bearing surface varied with each tumbler; a wide tumbler required a correspondingly deep notch in the key, a narrow tumbler required a shallow notch. When the key and tumbler stack matched, all tumblers would move in unison and release the bolt.

Barron's patent involved the key-bearing surface on the tumbler. Earlier locksmiths were content with a simple notch on the lower edge of the tumbler stack. When the tumblers were raised high enough, the bolt would release. Barron pierced the tumbler stack so that the key controlled both up and down movement of the individual tumblers. This refinement worked in conjunction with a more complex gating; the bolt would remain latched until all tumblers were at the same height. Because the tumblers move in two directions, this lock is described as a *double-action* type. Barron added up to six of these double-lever actions to his lock and thought it virtually impossible to open except with the proper key. He soon found out differently.

Another Englishman, Joseph Bramah, wrote *A Dissertation on the Construction of Locks,* which exposed the many weaknesses of existing so-called thiefproof locks and pointed out that any of them could be picked by a good lock specialist or criminal with some training in locks and keys. Bramah admitted that Barron's lock had many good points, but he also revealed its major fault: the levers, when in the locked position, gave away the lock's secret. The levers had uneven edges at the bottom; thus, a key coated with wax could be inserted into the lock and a new key could be made by filing where the wax had been pressed down or scraped away. Several trys could create a key that matched the lock. Bramah pointed out that the bottom edges of the levers showed exactly the depths the new key should be cut in order to clear the bolt. Bramah suggested that the lever bottoms should have a smooth surface and that the lever slots should be cut unevenly. Then only a master locksmith could open the lock.

Bramah's lock, patented in 1784, employed a series of notched lever tumblers that were aligned by corresponding notches in the key. The novelty was in the way the levers were arranged. Earlier locks were built on the pattern shown in Fig. 1-3; Bramah's lever tumblers were mounted vertically. Bramah used radial tumblers and a barrel key (Fig. 1-4). The pin on the side of the key rotated the lock.

Though this lock could be picked, the job was beyond the average thief. Each tumbler had to be aligned and the control piece had to be turned in the right direction.

Jeremiah Chubb added refinements to the Barron lock in an attempt to make it more secure. One of Chubb's improvements was a metal "curtain" which fell across the keyhole when the mechanism began to turn, making the lock difficult to pick. He also added a detector lever that

Fig. 1-3. An 18th-century lever-tumbler mechanism. The width of the tumblers corresponded to slots in the key.

Fig. 1-4. A Bramah radial lever lock (circa 1790).

indicated whether the lock had been tampered with. A pick or an improperly cut key would raise one of the levers too high for the bolt gate. This movement engaged a pin that locked the detector lever. The detector could be cleared by turning the correct key backwards and then forwards.

During this period, many robberies were committed against property and person. In 1817, the Portsmouth, England dockyard was robbed and the British Crown offered a reward to anyone who could devise a lock that could be opened *only* by its own key—one that was impossible to open by lockpicking or the use of a false key. A year later Chubb patented his lock and won the prize money.

This lock got much attention. One convict, a former locksmith, claimed he could open the best of locks. He was guaranteed $250 and his freedom if he could open Chubb's lock. The British Government and Chubb supplied the convict with a lock, key blanks, appropriate lockpicking implements, and some dismantled locks to work on. After working for some three months, he finally

gave up and served out his sentence. England announced that it finally had an unpickable lock.

Hobbs, an American locksmith, made a kind of minor career picking English locks. The manufacturers responded by developing new locks and manufacturing techniques. Many locks—over 3000—were patented in England alone during the 18th and 19th centuries. Key designs were as varied and ingenious as the locks they fitted. Perhaps the most interesting was the detachable bit key. The bits—the part of the key that worked the lock—could be disassembled and rearranged in different combinations. Only the owner knew the proper combination, if the key were lost or stolen it would be useless unless bitted correctly. Other keys had projecting pins like the Bramah key, and were intended for rotating locks. Others were flat with a wide tip, or socketed for detachable bits. Another type had a detachable tip. Elaborately notched keys were not uncommon, and were used with extremely complex locking wards. Each ward had a distinctive shape location in the lock. A false key might turn the lock a few degrees, only to be frustrated by additional wards.

The swivel key was another example of 19th century ingenuity. The lever, or the biting portion of the key, and the barrel, or shank, were moved independently of each other. Each half was turned in opposite directions to open the lock.

AMERICAN LOCKS

Early American lock bolts were mounted on the inside of the door, and could be opened from the outside by means of a latchstring. Hence the phrase, "the latchstring's always out." At night the string would be pulled inside, "locking" the door. Of course, someone had to be inside to release the bolt. An empty house was left unlocked. As the country was settled, theft increased; local merchants and blacksmiths soon got into the lock trade. Small cabinet and cupboard locks were very popular. The lady of the house (or, in large establishments, the butler) was in charge of the household keys. The doors were unlocked in the morning and the cabinets and cupboards were opened to dispense the day's supplies and immediately relocked. At night the doors were secured.

In the 1850s two inventors, Andrews and Newell, were granted patents on an important new feature—removable tumblers. The tumblers could be disassembled and scrambled to make, in effect, a new lock. The keys had interchangeable bits that matched the various tumbler arrangements. After locking up for the night, a prudent owner would scramble the key bits. Even if a thief got possession of the key, it would take him hours to stumble on the right combination. In addition to removable tumblers, this lock featured a double set of internal levers.

Newell was so proud of this lock that he offered a reward of $500 to anyone who could open it. A master mechanic took him up on the offer and collected the money. This experience convinced Newell that the only secure lock would have its internals sealed off from view. Ultimately, the sealed locks appeared on bank safes in the form of combination locks.

As we have seen, Hobbs picked the famed English locks with ease. Until Hobbs' time, locks were opened by making a series of false keys. If the series was complete, one of the false keys would match the original. Of course, this procedure took time. Thousands of hours might pass before the right combination was found. Hobbs depended upon manual dexterity. He applied pressure on the bolt while manipulating one lever at a time with a small pick inserted through the keyhole. As each lever tumbler unlatched, the bolt moved a hundredth of an inch or so.

Until the early 19th century, locks were made by hand. Each locksmith had his own ideas about the type of mechanism—the number of lever tumblers, wards, and internal cams to put into a given lock. Keys showed the same individuality. A lock could have 20 levers and weigh as much as five pounds.

Linius Yale changed all this. An inventor of milling machinery, Yale devised a simple, safe, and compact key-operated lock that could be made on automatic machinery. This lock was based upon the principle first developed by the Egyptians. Yale improved upon this principle and developed a keying system that is still used today. His son continued in the trade, making further refinements on his father's work. He developed a simplified cylinder with a rotating internal plug core that advanced lock design to new heights. Other refinements included the solid case and the small keyway that made life difficult for pick artists. The son also developed a series of dies and cutters to mass produce all parts of the lock.

The major structural difference between the Yale lock and the Egyptian prototype was the revolving plug that replaced the sliding bolt. A tailpiece on the back of the plug worked the bolt mechanism. Using machine tools, the Yale family was able to produce these locks with the same dimensions and tolerances. The keyway

Fig. 1-5. The operation of the Yale cylinder lock. The key raises the pin tumblers to the shear line and allows the plug to rotate.

was ridged to accept a grooved key blank. Only the correctly cut key would turn the plug and open the lock. This lock meant that thieves had to start over and devise new methods and tools. Figure 1-5 illustrates the basic Yale mechanism.

Competitors were quick to copy the Yale cylinder. The reliability of these locks revolutionized the industry. Some of the early lock companies are still in business.

Chapter 2

Tools of the Trade

The proper tools and equipment are crucial to the lock-smithing trade. How much money you spend on tools and equipment depends upon the amount of effort you are willing to put forth to build your own and the level of quality that you will accept in factory-made goods.

Consider some basic requirements. You will need tools and supplies to put the information you learn from this book into practice. The only way you can learn the trade is through practice. Be realistic in the assessment of what you will make, purchase locally, or purchase through locksmith supply houses. (A listing of various suppliers is provided in Appendix B.)

Space is another requirement. You should have a place free from distractions where you can study this book, store your supplies and equipment, develop projects based on the materials presented here, and still have room to be comfortable.

WORKBENCH

The workbench should be:

☐ Long enough to insure adequate work space.
☐ Strong enough to support the key machine at one end, out of the way.

☐ Solid enough to keep the machine in alignment.
☐ High enough so you can work without stooping.
☐ Wide enough to temporarily store parts—30″ is the comfortable maximum.
☐ Lit from overhead and behind or from the sides. Never have the light too close to the bench.

The location of the workbench must also be considered. If possible, it should be located in a well ventilated area, away from general traffic, near stored equipment and supplies, and with easy access from several directions. Bench storage facilities should be arranged to give instant access to frequently used tools. In this regard, a tool clipboard against the back of the bench is an excellent feature. Space for storage boxes, bins, and a shelf for reference books and manuals must also be considered (Fig. 2-1). It does no good to have the bench at one end of a room and all supplies and manuals at the other end.

Examples of workbench designs are shown in Figs. 2-2 and 2-3. No one example is ideal for everybody; individual tastes and needs differ.

A machinist's bench should be installed at one end, out of the way. It should be mounted on a swivel base, be sturdily built, and have jaws at least 3″ wide. The jaws should open to at least 5″ (Fig. 2-4).

Fig. 2-1. Arco bins are made of heavy polypropylene for long, maintenance-free service.

TOOLS AND SUPPLIES

Common tools—hammers, screwdrivers, etc.—are not discussed in detail here, but virtually all the tools and supplies used in locksmithing are included in the following lists. The most commonly accepted nomenclature is used in this book, although alternate names are given when applicable. The lists are not all-inclusive, but will more than suffice for learning the basics of locksmithing and starting a business. You may have some of the pieces of equipment at home, or you can use substitutes until you can afford professional equipment.

Necessary Tools

☐ Ace lock tools.
☐ Assorted springs.
☐ Automotive entrance tools.
☐ Brace and auger or expansion bits (Fig. 2-5).
☐ Brazing equipment.
☐ Broken key extractors.
☐ C-clamps (Fig. 2-6).
☐ Cam assortment.
☐ Cam screw assortment.
☐ Circular hole cutter.
☐ Code books.

Fig. 2-2. This simple bench is adequate for a student.

Fig. 2-3. This bench with its full complement of drawers and partitioned overhead bin is ideal for the professional.

Fig. 2-4. A Wilton professional-quality vise.

Fig. 2-5. You will need a good brace and collection of auger bits. (courtesy Stanley Works)

- ☐ Coping saw and blades.
- ☐ Cut-all knives and blades.
- ☐ Cylinder pin assortment.
- ☐ Depth gauge.
- ☐ Depth key assortment.
- ☐ Drills (hand and electric) and drill bits (Fig. 2-7).
- ☐ Files (warding, round, pippin, and triangular).
- ☐ Flashlight.
- ☐ Flat spring steel stock.
- ☐ Grinder with wire wheel.
- ☐ Hacksaw and blades (Fig. 2-8).
- ☐ Hammers (ball-peen and claw hammers are shown in Fig. 2-9).

Fig. 2-6. Stanley C-clamps are light and durable.

Fig. 2-7. Spade-type bits are used with power tools. (courtesy Stanley Works)

- [ ] Handcleaner.
- [ ] Hollow mill set rivet drills.
- [ ] Key blanks (assorted).
- [ ] Key caliper (with cutter assortment).
- [ ] Key clamp.
- [ ] Key layout board.
- [ ] Key machine.
- [ ] Lever tumbler assortment.
- [ ] Lock-aid tool.
- [ ] Lock pick set.
- [ ] Lock-reading tool.
- [ ] Lubricants (graphite, WD-40, and cup grease).
- [ ] Micrometer (Fig. 2-10).
- [ ] Piano wire.
- [ ] Pin tray.
- [ ] Pliers (adjustable, cutting, and locking lever).
- [ ] Plug holder.
- [ ] Plug followers (Fig. 2-11).
- [ ] Portable tool kit.
- [ ] Punch and die set in single and Double D configurations (Fig. 2-12).
- [ ] Punches (Fig. 2-13).
- [ ] Retainer ring assortment.
- [ ] Rim cylinder removal tools.
- [ ] Rivet assortment.
- [ ] Sandpaper and emery cloth.
- [ ] Screwdrivers (Fig. 2-14).
- [ ] Scribers.
- [ ] Shims.
- [ ] Sidebar cylinders and tumblers (automotive).
- [ ] Spindle assortment.
- [ ] Storage trays (Fig. 2-15).
- [ ] Tap set.
- [ ] Turning wrench.
- [ ] Tweezers.
- [ ] Vise.
- [ ] Wafer tumblers (assorted).
- [ ] Wood chisel set (Fig. 2-16).
- [ ] Wrenches (adjustable and pipe).

Useful Items

- [ ] Assorted nails and screws.
- [ ] Cigar boxes.
- [ ] Coat hangers.
- [ ] Ice trays.
- [ ] Identification tags.
- [ ] Masking tape.
- [ ] Matches and candles.
- [ ] Paper clips.
- [ ] Plastic glue.
- [ ] Plastic sandwich bags.
- [ ] Rubber bands.
- [ ] Soldering iron.
- [ ] Squeeze tube.
- [ ] String.
- [ ] Wood wedges.

TOOL SPECIFICS

The use of most tools is obvious and needs no discussion. However, a few are worthy of additional comment.

Broken key extractor—Used to extract broken keys from cylinder plugs.

Cup grease—Used only to lubricate the boltwork on exterior door locks, mortise cylinder locks, etc.

Fig. 2-8. The Stanley hacksaw features a tubular frame for maximum rigidity.

Fig. 2-9. Since a locksmith does both metalwork and simple carpentry, he needs both ball-peen and claw hammers. (courtesy Stanley Works)

Fig. 2-11. Plug tools. Insert a plug follower into the cylinder chamber as the plug is withdrawn. The follower prevents the pins and springs from falling into the chamber. Use the appropriate diameter plug holder as a vise.

Double D punch and die set—Used for cutting the lock mounting hole in a metal desk drawer.

Files—Various types are needed, including 8″ and 6″ mill, round (Swiss, fine tooth), 6″ triangular, 4″ flat warding, 6″ flat warding, and 6″ pippin. These files are used to cut keys by hand and are handy for miscellaneous repair jobs.

Graphite gun—Used to force powdered graphite into the tumbler mechanisms of pin and disc locks. (Oil and grease must not be used since these lubricants attract dirt.)

Hollow mill set rivet drill—Used for removing rivets from padlocks for access to the internal mechanism.

Key caliper—Used for measuring depths of key cuts; a caliper is especially useful when duplicating bit keys by hand. The caliper is set to the depth of the original key cuts and serves as a guide when filing the new key.

Layout board—Used to keep track of the pins and springs as they are removed from a lock plug and cylinder. It is important to reassemble these parts in their proper positions, otherwise the lock will have to

be rekeyed. A layout board is also used to position master key plug setups prior to assembly.

Machine oil—Used to coat the tumblers of lever locks. It is used as a lubricant and rust preventive on outside locks.

Pin tray—Used for holding various pins not necessarily ready for use in a specific lock. Many pin trays are arranged to carry an entire series of pins.

Plug followers—Used to secure the upper pins and springs of cylinder locks. The follower is inserted into the plug chamber as the plug is withdrawn. Popular chamber dimensions are given in Appendix C.

Thimble—Used to hold the plug when repairing it. This tool is also called a plug holder.

Tweezers—Used to handle small lock parts, especially the pins, during or in conjunction with disassembly or "setting up" of a pin-tumbler lock.

KEY MACHINE

A good key-duplicating machine is an expensive tool but one that you will need if you open your own shop. In

Fig. 2-10. Key micrometers are available from locksmith supply houses.

Fig. 2-12. A double-D punch makes a neat job of installing locks in metal drawer fronts.

Fig. 2-13. Purchase a good set of pin and center punches. (courtesy Stanley Works)

addition to duplicating cylinder and flat keys, the more sophisticated machines will reproduce warded and end-cut (Ace) keys of the type used on vending machines.

A key machine consists of three basic parts:

☐ A pair of vises coupled together and moving in unison. One vise holds the original key and the other holds the key blank.
☐ A key guide that reproduces the profile of the original key on the blank.
☐ A cutter wheel that notches the key blank on orders from the key guide.

Other features are useful but by no means absolutely essential. Some machines will duplicate a wide variety of keys with only a change in cutter wheel widths. Less sophisticated models require that the key holding vises be changed as well.

Critical Design Features

When evaluating a key machine, look carefully at the

Fig. 2-14. Screwdrivers. Electrician's screwdrivers (top) are designed for precision work; standard screwdrivers are used on the heavier jobs (bottom). (courtesy Stanley Works)

Fig. 2-15. Parts cabinets and storage boxes are a necessity. (courtesy Raaco)

Fig. 2-16. A Stanley wood chisel set.

Fig. 2-18. The Ket-Set KD-50, probably the most advanced and versatile key cutting machine available to the locksmith. (courtesy Taylor Lock Company)

pivot mechanism. The key must meet the cutter wheel squarely on a dead parallel with the axis of the wheel. A slight angle is enough to upset the dimensions of the duplicated key (Fig. 2-17).

Some machines don't have pivots and arrange matters so that the vises move laterally into the cutter. As long as the bearings are true, this ensures that the duplicated key is a faithful copy of the original.

The jaws of the vises must be carefully engineered to insure that both the original and blank key do not shift during the duplicating process, and that the keys are held square against the cutter. Some keys have groove dimen-

Fig. 2-17. A key machine should be constructed so that the key is almost parallel to the cutter. A long pivot (shown on the right) reduces the angle.

sions that make them difficult to grip. A piece of piano wire or a straightened paper clip can be used as a shim.

The Jet-Set KD-50 Deluxe is the ultimate key machine. It is considered by many to be one of the most advanced key machines in the world (Fig. 2-18). Superbly engineered, it is precisely machined and built for speed, accuracy, and profitability for the locksmith. Externally easy to operate, the machine will let you duplicate a key within five seconds. An automatic power-on, spring-loaded carriage and flip lever operation assure fast and accurate cuts. It also features a micrometer adjustable key guide, a lighted interior to provide proper visibility in key alignment, indexing vise jaw handles, and reversible heavy-duty two-way gripping vises. These vises are designed to let you cut almost any cylinder key, including single or double shoulder keys, keys without shoulders, narrow blade keys, double-sided keys (such as the Schlage wafer and many foreign car keys), and even the long, large bolt hotel keys. After clamping the keys into the vise jaws, a simple lateral movement of the flip lever (right and then left) is all that it takes to cut the duplicate key.

Cutters

There are three basic types of cutters. *Filing* cutters and *milling* cutters are the most popular. A *slotter* cutter is used on square-ended keys and to make side cuts on keys for lever and warded locks. Filing and most milling cutters are flat on one side and beveled on the other (Fig. 2-19). The flat side makes the right angle cut that is usually required next to the shoulder of the key. These cutters are always mounted with the flat side towards the original key.

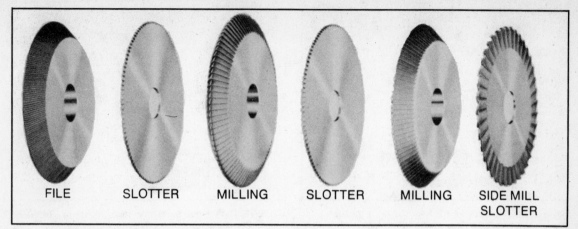

FILE   SLOTTER   MILLING   SLOTTER   MILLING   SIDE MILL SLOTTER

Fig. 2-19. Various cutter wheels.

The diameter of the cutter is also important. Large diameter cutters leave a very slight, almost imperceptible, concave on the key blank. Smaller cutters make a deep concave in the key. The greater the concavity, the more rapidly the key wears.

Key Guide

The *key guide* should be checked occasionally. Mount two identical key blanks in the vise and rotate the cutter by hand. If all is correct, the cutter wheel should just touch the right-hand key.

Figure 2-20 shows the adjustment procedure used on Keynote Engineering's M-76 machine. Most other makes use a similar setup. Loosen nut A and, holding the guide edge perpendicular, turn nut B in or out. Tighten nut A and test. Several tries are usually required to get the adjustment right.

HOLE SAWS

As a locksmith, you're going to be installing a lot of locksets in both commercial establishments and residences. You want efficiency plus the time-saving prod-

Fig. 2-20. The cutter wheel should barely touch the key blank (left); if adjustment is required, loosen nut A and move nut B in or out. (courtesy Keynote Engineering)

15

Fig. 2-21. Locksmith's kit for hole cutting, a necessity for greater efficiency and time-saving in the locksmith's work. (courtesy The M. K. Morse Company)

Fig. 2-22. High speed steel hole saws. (courtesy The M. K. Morse Company)

ucts that will do the various jobs required. Among these products are pilot drills and hole saws. The M. K. Morse Company has developed an excellent line of these, and has especially considered the locksmith in this development. Among the best products for the locksmith's consideration is the Morse Locksmith Kit (Fig. 2-21). A self-contained functional set of the basic high-speed steel hole saws with a built-in arbor, the kit contains eight different sizes (from ¾″ to 2⅛″) to meet varying needs, plus four pilot drills and a screwdriver.

The high-speed steel hole saws have welded edge construction, and they saw holes in wood, metal, or any machinable material. The great advantage of this kit is that it has been specially designed for the locksmith. Each hole saw has a 1⅞″ cutting depth, and will saw through standard doors or 2×4s in one pass.

Fig. 2-23. Hole saw kit. (courtesy The M. K. Morse Company)

Fig. 2-24. A variety of arbors complete with pilot drills. (courtesy The M. K. Morse Company)

Many locksmiths have several of these kits for use, keeping one in the van, one for each employee, and one in the shop. In addition, it is not unusual to be doubly prepared and have several extra units (Fig. 2-22) and individual hole saw arbors available. Sometimes, for large volume and special applications, a general hole saw kit is also available and put to numerous uses (Fig. 2-23).

A variety of arbors are available and must also be considered for a complete shop equipment layout (Fig. 2-24). These arbors (complete with pilot drills) are for ¼" high-speed drills, although a universal hole saw arbor for ½" drills is also available.

Each hole saw is packed at the factory with a table showing the recommended rpm for each size saw cutting specific materials. Operating at higher speeds than those recommended will shorten the life of the saw and produce some inefficient cutting. You save the cost of a new hole

saw and increase your production efficiency by ensuring that the operating speed is correct in every case.

Sufficient feed pressure to take a chip when using the hole saw is necessary. Variables in materials, work configuration, etc., should be considered. A rule of thumb is to apply 80-100 pounds per inch or hole saw diameter when sawing in metals. Insufficient feed pressure will dull the tooth points prematurely, while excessive pressure can destroy the teeth.

Cutting oil serves two purposes when sawing in metals. It cools the saw and the work, and it also removes chips from the kerf. Cutting oil therefore reduces heat and abrasion which can shorten the cutting life.

Let's now look at two of the hole saws that you should be using in your locksmithing shop. The long life deep cutting saw (Fig. 2-25) combines the convenient built-in arbor with another innovation—slots in the blade

Fig. 2-25. Long life deep cutting hole saw. (courtesy The M. K. Morse Company)

Fig. 2-26. Carbide-tipped hole saw. (courtesy The M. K. Morse Company)

17

M44K STEM

3 PINS THIS SIDE

M44K PIN PLATE

2 PINS THIS SIDE

M44K NUT

Fig. 2-27. Arbor construction and parts relationship. (courtesy The M. K. Morse Company)

body. The slots allow greater chip clearance and reduce the heat buildup for faster, easier cutting and longer unit life. The 1⅞" depth for one pass hole sawing in doors and 2" dimensional lumber is an economical and time-saving advantage, and also a necessity in many industrial applications for which you may be called upon.

The second unit is the carbide-tipped hole saw (Fig. 2-26) for abrasive materials such as fiberglass, particle board, and asbestos board. It has three teeth per inch and is carbide-tipped to resist abrasion. It is not recommended for ferrous metals, but will outlast high-speed steel hole saws in applicable jobs.

The speed of operation of high-speed steel hole saws is critical: a 1" hole saw, for example, should be operated at 350 rpm in mild steel. For other sizes of hole saws, divide the rpm by the hole saw diameter in inches.

A good grade of cutting oil should be used.

The point of the pilot drill must not exceed beyond the teeth points more than the thickness of the material

EXTEND PILOT DRILL THRU M44K STEM FOR ¼" CHUCK

M44K STEM FOR SMALL ½-20 THREAD HOLE SAWS

M44K ARBOR FOR BLACK & DECKER MANDREL #40092 WITH 3 PIN DRIVE (Reverse M44K Pin Plate)

USE N-1 or N-2 ADAPTOR NUTS FOR BLACK & DECKER MANDRELS AS SHOWN:

N-1 NUT

N-2 NUT

#40090 BLACK & DECKER MANDREL
Place N-1 Nut on the M44K Stem.
N-1 NUT

#40091 BLACK & DECKER MANDREL
Place N-2 Nut on the M44K Stem.
N-2 NUT

#40093 BLACK & DECKER MANDREL
Extend Pilot Drill thru M44K Stem for ¼" chuck. Place N-1 Nut on M44K Stem.
N-1 NUT
PILOT DRILL

#40094 BLACK & DECKER MANDREL
Extend Pilot Drill thru M44K Stem for ¼" chuck. Place N-2 Nut on M44K Stem.
N-2 NUT
PILOT DRILL

Fig. 2-28. Proper alignment and various attachments of arbor; precision is necessary for ensuring the proper use of the hole saw. (courtesy The M. K. Morse Company)

being cut. When the drill pierces the material, the teeth must already be engaged.

The drill, arbor, and hole saw combination must be held perpendicular to the surface of the material being cut.

Feed pressure should be about 30 pounds for each inch of circumference in contact with the work (five pounds per tooth).

Let's assume for the moment that you have a locksmith's hole saw kit but it is unavailable, so you go to the workbench shelf and get the hole saw arbor for your ½″ Black and Decker drill; you also get a steel hole saw to attach to it. Be sure you follow directions on how to properly attach the two together. Figure 2-27 shows the construction and positioning interrelationships of the various parts of the arbor (Morse model M44K). Figure 2-28 gives information relative to ensuring the arbor and hole saw are properly aligned and attached before use.

LOCKS FOR STUDY

A beginning locksmith needs all kinds of locks to examine, take apart, repair, and cannibalize for parts. Second-hand locks are available from pawn shops, auto wrecking yards, and abandoned buildings. You must have permission to enter buildings that are to come under the wrecker's ball. Check with the superintendent or, if demolition has already begun, with the job foreman. Office buildings are a particularly rich source of locks. A search of the storage areas may turn up lock parts and spare keys.

Chapter 3

Types of Locks and Keys

The locking devices commonly used today can be grouped into a number of general types. These types are:

☐ Warded locks.
☐ Mortise locks.
☐ Automobile locks.
☐ Key-in-knob locks.
☐ Surface-mounted auxiliary locks.
☐ Lever locks.
☐ Luggage locks.
☐ Padlocks.
☐ Combination locks.

An example of each type is shown in Fig. 3-1. Within these few major types all design variations fall. Some locks, such as a combination door lock, represent a melding of two different types.

In this chapter we will consider the various types of locks and keys that are in common use today (Fig. 3-2).

Warded locks are either of the *rim* or *mortise* type and are among the oldest locks still being made (Figs. 3-3 and 3-4).

Lever and *wafer* locks are related, although they were introduced to the public at different times.

A variety of *padlocks* exist—warded, wafer, disc, and pin tumbler. Smaller padlocks have a single ward and take very simple keys. The common railroad lock is a padlock. It is opened with an old-style barrel key and uses a simple lever mechanism. Some railroad locks work dependably even after being exposed to the elements for fifty years. Padlocks can also be combination locks.

The simplicity of the cylinder lock made possible its mass production. The cylinder is a single rotating unit firmly encased within a solid metal housing. The creation of the Yale cylinder lock encouraged several variations: the key-in-knob lock, the double lockset (Fig. 3-5), and the desk lock. A variation of the cylinder lock used in coin-operated machines is the Ace circular lock (Fig. 3-6).

A combination lock consists of a series of interconnecting wheels rotating about a central core that is controlled in its revolutions by an outside combination knob. The number of revolutions of the knob necessary to open the lock is pretty much standard: a 3-2-1 sequence is most common; some high-security locks may require a 4-3-2-1 rotation. Combination locks are of two basic types: the hand and the key-change variety. A kind of hand combination lock not dependent upon internal wheels is the pushbutton lock, shown in Fig. 3-7.

Fig. 3-1. The nine basic locking devices.

Fig. 3-2. Four basic key types in use today.

Fig. 3-3. Warded-key cuts.

Fig. 3-4. Typical warded locks. (courtesy Taylor Lock Company)

Surface-mounted auxiliary locks include dead-latches, surface-mounted cylinders (when used separately from another lock unit), chain door guards, rim latches, surface bolts, and chain bolts (Fig. 3-8).

The automotive lock employs the sidebar principle. When the key engages the tumblers, it aligns them so that a sidebar built into the lock drops into place, allowing the key to turn to open the lock. The sidebar principle is also used on certain pin-tumbler cylinders.

KEY BLANK IDENTIFICATION

Three parts of the key blank are used to match a given key with a key blank:

☐ The *head shape* of the key.
☐ The *length* of the key.
☐ The *millings* (grooves) of the key.

The head shape will usually pinpoint the manufacturer, since most manufacturers like to give their keys an individual look. The length and millings are very critical. To determine the length, lay the key over the illustration in the key catalog and align the shoulders. The length is then observed at a glance because the key catalog drawings are the size of the original keys. To match the grooves, stand the key directly over the cross-section that appears under each blank. The grooves must match *exactly* to be considered the same.

Key manufacturers put out catalogs that identify their key blanks, their sizes, and a cross section of each different blank. In addition, they may have a cross-reference section which refers to other manufacturers' keys that are applicable to each of the keys in that particular catalog. These cross-reference sections are valuable as they obviate reference to a wide variety of catalogs in order to determine the manufacturer of a given key blank. Merely

Fig. 3-5. A double lockset. (courtesy P & F Corbin)

STUD

COUPLER

CASE

OUTER BUSHING

TUMBLER SPRING

SPRING PIN

INNER BUSHING

TOP PIN

BOTTOM PIN

Fig. 3-6. A cross section of an Ace circular lock. (courtesy Desert Publications)

refer to the key type and cross-reference it to the one that you have.

A problem that arises here is that some manufacturers have literally *hundreds* of key blanks, and the cross-reference section becomes unwieldly at times. A catalog that carries the most popular and commonly used blanks is best to use. About 95 percent of the keys you will have to duplicate will be found in such catalogs. Other catalogs can be used for general reference for the other 5 percent, but are not absolutely necessary to have for your shop.

KEY BLANK EXAMPLES

On the following pages are illustrations of various types of key blanks. Figure 3-9 shows key blanks from the Taylor Lock Company. These cylinder blanks will fit a variety of American locks. Notice that in Fig. 3-10 the blank is further laid out to incorporate a master key system and from the cross section, you can identify the various keyways that will fall into this particular master key system.

Figure 3-11 shows another keyway system with the applicable blanks. Notice that within this key series there are three different key lengths, which allows for a wider use of the master key system. The length also allows for more pins to be used in the cylinders. This means a greater variety in the number of possible key combinations available to the system.

Foreign automobiles (and, naturally, their keys) are increasing in popularity, so it is necessary to have a

Fig. 3-7. A Simplex pushbutton combination lock.

23

RIM LATCH

DEADLATCH

SURFACE BOLTS

SURFACE-
MOUNTED
CYLINDER

CHAIN
BOLT

CHAIN DOOR GUARD

Fig. 3-8. Surface-mounted auxiliary locks.

well-stocked variety of foreign lock keys in the locksmith shop. Figure 3-12 provides illustrative details on a variety of the more common key blanks in use. You should have these types, and also a slightly wider supply of other foreign auto key blanks available to service the needs of your customers.

Automotive Key Blanks

The duplication of automotive keys is more prevalent in recreational areas and very large cities. Lost or stolen keys mean duplication and/or rekeying on a weekly basis for many locksmiths. It becomes imperative that as a locksmith you have in your shop an ample and wide selection of key blanks on hand in this particular area. Keep a selection of the most common and popular automotive blanks in your van plus one or two others that might be required. (Note: A standard selection of automotive key blanks should be in *every* locksmith shop, but in

large cities and in recreational areas such as beaches, parks, etc., the need is greater than usual. For this reason, in these areas, the number and variety of blanks will be greater than in others. Always consult with your distributor or manufacturer's representative to obtain a more accurate view of the types and quantities of blanks that are necessary; don't buy a larger selection than required just because you think you may *some day* need it.

Figure 3-13 is a breakout of domestic automotive blanks, identified by the number and the vehicle they accompany. These are, by far, the most common ones for which you will be cutting keys for. (This is a representational selection of popular keys that require frequent duplication; it is not all-inclusive.) Refer to your key catalog for illustrations of all the automotive keys and start to develop a working memory of them. You should be able after a period of time to look at a key that is already cut (or even a blank) and be able to determine

Fig. 3-9. Representative sampling of some standard key blanks. (courtesy Taylor Lock Company)

Fig. 3-10. Six pin cylinder key and various associated keyways. Keyway M is used only for the master key. (courtesy Taylor Lock Company)

with relative accuracy what make of automobile the key goes to and where the key blanks are located on your key blank board. Figure 3-14 shows some of these key blanks. Figure 3-15 lists the most common foreign keys by number of automotive type, while Fig. 3-16 shows a variety of these keys.

Look-Alike Keys

As the name implies, you will find key blanks that are "look-alike" keys—blanks made by a manufacturer other than the original that look just like the original blank. These provide you with a great advantage; as look-alikes, they are quickly identified with a specific lock manufacturer, and you can thus quickly set the proper key for duplication.

Not all keys made are look-alikes. This brings up an important point—comparison charts or cross-reference key charts. They go by either name but the function is the same, and they are an important reference tool in your shop. You may have a customer who brings in a key for

To order keys, keyway section is added to pin length. Example - 43 LA is 5 pin length, having LA keyway and A43LH is submaster in 7 pin length to by pass LE, LF and LG key sections.

LN = 49 KM (5 pin)
      O49 KM (6 pin)
      A49 KM (7 pin)

43 Series
5 pin

O43 Series
6 pin

A43 Series
7 pin

Fig. 3-11. Larger master key system which provides a greater variety of key uses and control features through sub-masters and individual use keys. (courtesy Taylor Lock Company)

Fig. 3-12. Variety of popular key blanks for foreign automobiles. (courtesy Taylor Lock Company)

AD1	Audi 100 and 100LS—1971 on	LU1	GM LUV Ign./Door & Trunk/Glove—1973 on
CP1	Capri Ign. 1971 on—Ign./Trunk 1975 on	MZ1	Mazda Door/Trunk—1971 on
CP2	Capri Door/Trunk 1971-74	MZ2	Mazda—Supplements MZ1
CP3	Capri—Supplements CP2	MZ3	Mazda, Courier Trucks 1971-73
CP4	Capri—Fiesta	MZ4	Mazda/Courier R100/1200-Door/Trunk 1970-72
5DA1	Datsun	MZ5	Mazda—Supplements MZ4
5DA2	Datsun 1970-72, Mazda 1971 on, Eng. 1969 on	RP1	Renault—Peugeot
DA3	Datsun—1970 on	5RP2	Renault-Peugeot Ignition 1971 on
DA4	Datsun & Subaru—1970 on	TO1	Toyota
LDC1	Long Head Dodge Colt, Arrow, Chall., Sapporo, Jap. Opel—1970 on	TO2	Toyota Ignition—1969 on
		UN2	Union (Hillman—Vauxhall—English Ford)
5FT1	Fiat Ignition—1967 on	UN3	English 1960 on, Volvo Door/Trunk 1970
FT2	Fiat Trunk 1967 on—Pantera-Ferrari	UN4	Union—MG—Triumph
FT3	Fiat—Supplements FT2	6VL1	Volvo 1960 to 1965-6 Pin
HN1	Honda Ignition 1970 on, MG Ignition 1972	VW1	Volkswagen
HN2-3	Honda MASTER (Codes #1001-1700, 2001-2700)—1976	VW2	Volkswagen 1965-70
		VW3	Volkswagen Beetle—1971 on
HN4	Honda Trunk 1973-76	VW4	Volkswagen VW411-412-DA-RA-SC, Fox, Audi 4000/5000, Porsche, Jetta
HN5	Honda Cycles—1977 on	VW5	Volkswagen Bus—1971 on

Fig. 3-13. Most popular foreign automotive keys listed by key number and automotive use. (courtesy Star Key and Lock Co.)

Fig. 3-14. Representative selection of commonly duplicated automobile keys. (courtesy Star Key and Lock Co.)

HBR1	American Motors Ignition/Door to 1959
OBR1	American Motors Trunk/Glove to 1959, GAS locks
OBR1DB	Briggs & Stratton—GAS & utility locks
HBR2	GM Ignition/Door—all years to 1966
OBR2	GM Trunk/Glove—all years to 1966
HBR3	American Motors Ignition/Door—1960 on
OBR3	American Motors Trunk/Glove—1960 on
OBR4	GM Long Head Ignition/Door—all years to 1966
HBR5	GM Ignition/Door "A"—1967
HBR5M	GM Ignition/Door Master "A/C"—1967/68
OBR5	GM Trunk/Glove "B"—1967
HBR6	GM Ign./Door & Trunk/Glove Master—hexagon head—1967
OBR6	GM Ign./Door & Trunk/Glove Master—oval head—1967
HBR7	GM Ignition/Door "C"—1968
OBR7	GM Trunk/Glove "D"—1968
HBR8	GM Long Head Ign./Door Master "A/C"—1967/68/71/72 (Ign. only 1975/76/79/80)
HBR9	GM Long Head Ign./Door & Trunk/Glove Master "E/H"—1969/73/77/81
OBR9	GM Oval Head Ign./Door & Trunk/Glove Master "E/H"—1969/73/77/81
HBR9E	GM Ign./Door "E"—1969/73 (Ign. only—1977/81)
OBR9H	GM Trunk/Glove "H"—1969/73 (Door also—1977/81)
HBR10J	GM Ign./Door "J"—1970 (Ign. only—1974/78/82)
OBR10K	GM Trunk/Glove "K"—1970 (Door also—1974/78/82)
HBR11	American Motors Ign./Door—1970 square head—1960 on
OBR11	American Motors Trunk/Glove—1970 ellipse head—1960 on
HBR12A	GM Ign./Door "A"—1967/71 (Ign. only—1975/79)
OBR12B	GM Trunk/Glove "B"—1967/71 (Door also—1975/79)
OBR13	GM Ign./Door & Trunk/Glove Master "A/B"—1967/71/75/79
OBR13S	GM Ign./Trunk—fits most years—1967 on—FOR LOCKSMITHS ONLY—NOT FOR DUPLICATION
HBR14C	GM Ign./Door "C"—1968/72 (Ign. only—1976/80)
OBR14D	GM Trunk/Glove "D"—1968/72 (Door also—1976/80)

HFD1	Ford Ignition/Door—1952 to 1954
OFD1	Ford Trunk/Glove—1952 to 1954
HFD2	Ford Ignition/Door—to 1951
OFD2	Ford Trunk/Glove—to 1951
HFD3	Ford Ignition/Door—1955 to 1958
OFD3	Ford Trunk/Glove—1955 to 1958
HFD4	Ford Ignition/Door—1952 to 1964
OFD4	Ford Trunk/Glove—1952 to 1964
HFD5	Mercury Ignition/Door—1952 to 1964
OFD5	Mercury Trunk/Glove—1952 to 1964
HFD6	Ford, Linc., Merc. Ign./Trunk Square Head Master—1952 to 1964
OFD6	Ford, Linc., Merc. Ign./Trunk Round Head Master—1952 to 1964
HFD7	Ford Ign./Door to 1951—Studebaker 1947 on
HFD8	Lincoln Ignition/Door—1952 to 1964
HFD9	Ford Ignition/Door—1965/66 only (double side)
OFD9	Ford Trunk/Glove—1965/66 only (double side)
HFD10	Ford Ignition/Door—1965 on (double side)
OFD10	Ford Trunk/Glove—1965 on (double side)
HPL1	Chrysler, Plymouth, Dodge Ign./Door 1949-55
OPL1	Chrysler, Plymouth, Dodge Trunk/Glove 1949-58
HPL2	Chrysler, Plymouth, Dodge Ign./Door to 1948
OPL2	Chrysler, Plymouth, Dodge Trunk/Glove to 1948
HPL3	Chrysler, Plymouth, Dodge Ign./Door 1956-59
OPL4	Chrysler, Dodge Trunk/Glove 1959-65
OPL5	Plymouth Trunk/Glove 1959-65
HPL6	Chrysler, Plymouth, Dodge Ign./Door 1960-67
OPL7	(Master) Long GM to 1966 & Chrysler Trunk 1959-65
OPL8	Chrysler Trunk/Glove—1966/67
HPL68	Chrysler Ign./Door Master 1956 on—hexagon head
OPL68	Chrysler Trunk/Glove Master—1966 on
OPL70	Chrysler Ign./Door & Trunk/Glove Master—1966 on
HPL73	Chrysler Ign./Door Master 1956 on—diamond head
HYA4	Kaiser Frazer Ign./Door—1949/50
HYA5	Studebaker Ign./Door—1949 on

Fig. 3-15. Automotive key blanks by key number and associated auto use. (courtesy Star Key and Lock Co.)

duplication, and for one reason or another, you cannot put your finger on the specific key blank that is required in order to duplicate it. Figure 3-17 is a manufacturers' key comparison chart for automotive and house keys.

From this chart you have, say, two keys that require duplication—one for the house and the second for the family car. The house key has a National/Curtis number of IN1, whereas the car key is 1127DP. Here's how to quickly determine whether you have the particular blanks for these two keys or not. (As an up-and-coming locksmith, you carry a wide assortment of the most commonly used keys, plus you have a cross-reference key chart listing.)

Go to the cross-reference list and find that the IN1 goes to the Star blank 5IL1 (Ilco house lock). The 1127DP is an Ilco key, for which you do have the Star blank, which

is HFD4, going to a Ford auto ignition. Rapidly looking at the key blanks on the way, you now quickly select the two proper keys. Within minutes you have another satisfied customer, thanks to the cross-reference listing.

The cross-reference listing is available from your key blank representative, or you can spend between $20 and $50 and get a complete cross-reference book. This is both an advantage and a disadvantage. With the full book, you may have to look in a number of sections, whereas with the smaller individual cross-reference breakout from your key distributor/manufacturer, you are more likely to find the required key almost instantly.

A copy of a cross-reference listing is a prime requisite for every shop. Remember, go first with the listing obtained from your key manufacturer; the book can come later, if it is ever really needed.

Fig. 3-16. Selection of popular foreign automotive key blanks that require frequent duplication. (courtesy Star Key and Lock Co.)

AUTOMOTIVE

STAR	Ilco	Natl/Curtis	Make
HBR2	H1098LA	B10	GM Ignition to 1966
OBR2	01098LA	B11	GM Trunk to 1966
HBR3	H1098NR	B24,RA2	Amer. Mtrs. Ign. 1960 on
OBR3	1098NR	B4,RA1	Amer. Mtrs. Trunk 1960 on
HBR5	H1098A	B40	GM Ignition "A"-1967
HBR9E	P1098E	B44	GM Ignition "E" - 1969 & 1973 (Ign. only-1977/81)
OBR9H	S1098H	B45	GM Trunk "H" - 1969 & 1973 (Door also-1977/81)
HBR10J	P1098J	B46	GM Ignition "J" - 1970 (Ignition only-1974/78/82)
OBR10K	S1098K	B47	GM Trunk "K" - 1970 (Door also-1974/78/82)
HBR11	1970AM	RA4	Amer. Mtrs. Ignition - 1970 square head - 1960 on
OBR11	S1970AM	RA3	Amer. Mtrs. Trunk - 1970 ellipse head - 1960 on
HBR12A	P1098A	B48	GM Ignition "A" - 1971 & 1967 (Ign. only-1975/79)
OBR12B	S1098B	B49	GM Trunk "B"-1971 & 1967 (Door also-1975/79)
HBR14C	P1098C	B50	GM Ignition "C" - 1968/72 (Ign. only-1976/80)
OBR14D	S1098D	B51	GM Trunk "D" - 1968/72 (Door also-1976/80)
HFD4	1127DP	H27	Ford Ignition 1952-64
OFD4	1127ES	H26	Ford Trunk 1952-64
HFD10	1167FD	H33,H51	Ford Ign.Double Side 1965 on
OFD10	S1167FD	H32,H50	Ford Trk.Double Side 1965 on
OPL4	1759L/1764S	Y141,B29-31	Chrysler Trunk 1959-65
HPL68	1768/69/70CH	Y152/150/148/146	Chrysler Ign.Master 1956 on
OPL68	S1768/69/70CH	Y151/149/138	Chrysler Trk.Master 1966 on

HOUSE

STAR	Ilco	Natl/Curtis	Make
5CO1	1001EN	CO7	Corbin X1-67-5
5DE3	D1054K	DE6	Dexter 67
5EA1	1014F,X1014F	EA27,EA50	Eagle 11929 - Harloc
5HO1	1170B	HO3/HO1	Hollymade - Challenger K1
5IL1	1054F	IN1	Ilco 1054F - Keil 159AA
5IL2	1054K	IN3	Ilco - Dexter - Weslock
5IL7	X1054F	IN21,IN18	Ilco X1054F
5KE1	1079B	K2	Keil 2KK
5KW1	1176	KW1,PT1	Kwikset 1-1063 - Donner - Petco
5LO1	1004	L1	Lockwood B308 - 5 Pin
MA1	1092	M1	Master 1K,77K
5RO4	R1064D	NA5, NA6	Rockford 411-31A
5RU2	1011P	RU4	Russwin 981B
5SA1	01010	S4	Sargent 265U
5SE1	1022	SE1	Segal K9 - Norwalk - Star
5SH1	1145	SC1	Schlage 35-100C(923C)
SH2	1307A	SC6	Schlage 35-180(920A)
SH6	1307W	SC22	Schlage 35-200(927W)
5TA4	1141GE	T4,T7	Taylor 111GE
5WK1	1175, 1175N	WK1/WK2	Weslock 4425
5WR2	1054WB	WR2,WR3	Weiser 1556 - Falcon
4YA1	999B	Y145,Y220	Yale 9½
5YA1	999	Y1	Yale 8 - Acrolock

Fig. 3-17. Key comparison chart for automotive and residential key blanks. (courtesy Star Key and Lock Co.)

Chapter 4

The Warded Lock

The warded lock is the oldest lock still in use and is found in all corners of the world. It employs a single or multiple warding system. Because of its simple design, its straightforward internal structure, and its easily duplicated key, this lock is an excellent training aid for locksmiths. This same simplicity means that warded locks give very little security. Use these locks only in low-risk applications such as storage sheds and rooms where absolute security is not essential.

At one time, warded locks were used on most doors. These locks are still found in abundance in older buildings still standing in decayed metropolitan neighborhoods, such as the center city of Philadelphia, Market Street in San Francisco, the Old Town section of Chicago, and the East Side of New York City.

The oldest of these buildings have the cast iron locks on the doors; some date back to the last century. Later types were made of medium gauge sheet metal. The casing consists of two stampings, the cover plate and the backplate. The latter mounts the internal mechanism and forms the sides.

The warded lock derives its name from the word *ward*, meaning *to guard*. The interior of the lock case has protruding ridges or wards that help to protect against the use of an unauthorized or improperly cut key. Normally, there are two interior wards positioned on the inside of the cover and backing plate, directly across from each other.

This lock is sometimes mistakenly referred to as a *skeleton-key* lock. The proper and full name is the *warded-bit key* lock.

TYPES

Two types of warded locks are currently in use, the *surface-mounted* or *rim* lock and the *mortised* lock (Fig. 4-1). While both types are similar in structure and size, the degree of security they give varies. The internal mechanisms of both operate on the same principle, but the mortise lock may have several additional parts. Differences between these locks are:

Surface-mounted (rim) lock	Mortised ward lock
Mounted on door surface.	Mounted inside of door.
Secured by screws in the doorface.	Secured by screws in the side of the door at the lock faceplate.

Surface-mounted (rim) lock	Mortised ward lock
Door can be any thickness.	Door must be thick enough to accommodate.
Thin case.	Fairly thick case.
Short latchbolt throw.	Up to 1″ latchbolt throw.
Lock from either side.	Locked from either side.
Strike can be removed with door closed.	Strike cannot be removed with door closed.
Very restricted range of key.	Restricted range of key changes.
Very weak security.	Weak security.

CONSTRUCTION

The basic interior mechanism is drawn in Fig. 4-2. Since the relative security of any lock lies in the type of key used, the number of key variations possible, and the amount of access to the locking mechanism afforded by the keyhole, the warded lock is the least secure.

If a lock were designed to have no more than 10 different key patterns (changes), and 1000 locks were made, 10 different keys would open all 1000 locks. By the same token, one key would open the lock it was sold with and 99 others. Furthermore, it is often possible to cut away parts of a key to pass (negotiate) the wards of all 1000 locks. From this you can see that lock security is related to the kind and number of key changes built into the system when it is initially designed.

The keyhole is an access route to the interior mechanism of the lock. The larger the keyhole, the easier it is to insert a pick or other tool and release the bolt.

In theory, each warded lock may be designed to accept 50 or even 100 slightly different keys. In practice, these locks tend to become more selective as they age and wear. The lock may respond to the original key or to one very much like it, but other keys that would have worked when the mechanism was new no longer fit.

While this may seem fine and well for the lock owner, excessive wear increases the potential of key breakage within the lock and of jambs in the open, partly open, or closed position. It can also mean that a new lock will have to be installed.

Most surface-mounted and mortised locks are intended to be operated from both sides of the door. Keyholes and doorknob spindle holes extend through both sides of the lock body. Occasionally you will en-

Fig. 4-1. Warded locks are available in mortised (left) and surface-mounted varieties. (courtesy Taylor Lock Co.)

Fig. 4-2. The internals of a typical warded lock.

Fig. 4-3. The key on the right is for a lock activated from one side of a door; the key on the left can pass the wards from either side.

keyhole through the door and lock body and may require some filing on the key. Figure 4-3 illustrates the differences in keys. Note the additional cut on the left-hand key.

OPERATION

The key must be cut to correspond to the single or multiple side and end wards that have been designed into the lock. After the key passes these wards, it comes into contact with the locking mechanism. The cuts on the key lift the lever to the correct height and throw the dead bolt into the locked or unlocked position.

Turning the doorknob activates the spindle and, so long as the dead bolt is retracted, releases the door.

Figure 4-4 depicts various keyhole control features that allow only certain types of cut keys to enter the keyhole. Figure 4-5 shows a key entering a keyhole. Notice that the key has the appropriate side groove to allow it to pass through the keyhole and into the lock. If you were to file off this obstruction (called a *case ward*), any key thin enough to pass could enter. (Some ward bit keys are quite thick.) By the same token, a very thin key is able to pass whether or not the side ward is present (Fig. 4-6). The common skeleton key is a prime example of this; it is thin enough to pass most case wards, but it will not necessarily open the lock!

Figure 4-7 shows a key engaging the bolt. While there is only one set of wards in this particular lock,

counter a surface-mounted lock with a doorknob spindle and keyhole only on one side. The other side is blanked off.

A lock of this type can be modified to accept a key from the other side. This modification entails cutting a

Fig. 4-4. Case ward variations.

Fig. 4-5. A slot milled on the edge of the key allows the key to pass the case ward.

because this ward is set where it is, it gives more security than a lock with no ward, or one with a ward that has been worn down to almost nothing.

REPAIR

Warded locks are not repaired to any great extent since replacement is usually cheaper. You should, how-ever, have a supply of spare parts for these locks. The most frequent failure is a broken spring. Over a period of time, the spring may crystallize where it mates with the bolt. In addition, the wards can break or wear down into uselessness.

Replace a broken spring with a piece of spring stock, cut to length and bent to the correct angle. Some spring stock must be tempered before use; other springs come already tempered. If tempering is necessary, heat the spring cherry red and quench it in oil. You can save time by purchasing standard springs already bent into a variety of shapes that are designed to fit almost all locks.

Worn or broken wards on locks with cast cases can be repaired by drilling a small hole in the case and forcing a short brass pin into the lock case. The best technique for brittle cases is to braze a piece of metal on the case at the appropriate spot and file it down to the appropriate size. Wards on locks with sheet metal cases can be renewed by indenting the case with a punch ground to a fairly sharp point. If the factory has already punched out the wards, it would be best to braze a piece of metal at the proper spot.

Since most of these locks are inexpensive and offer minimal security, you should remind the customer that the cost of repair may far exceed the cost of the lock. Purchasing a new and more secure lock has definite advantages, and since you are already there, the homeowner's cost would then be less than ordering one and having you make a second trip to install it.

Should the homeowner decide to take your advice and purchase a new lock, ask to keep the broken one; it is of no use to him and parts are always nice to have. Sooner or later you will have to repair another lock of the same type; having the correct parts at hand will save you time. Furthermore, you have made a sale, and by being allowed to keep the old lock, have obtained parts at no cost.

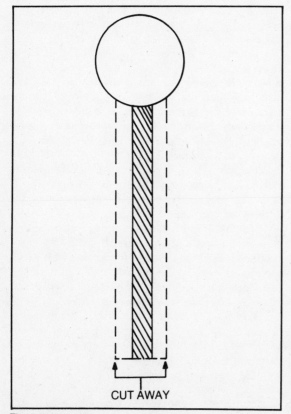

Fig. 4-6. A skeleton key is one that has been ground down to defeat the case ward.

Fig. 4-7. The single ward is better than none.

Removing paint from these locks is also a form of repair. A major cause of lock failure is the home painter. He does not take the time to remove the lock or to cover it with masking tape. Some paint usually gets into the mechanism. To clean a paint-bound lock, follow this procedure:

Remove the lock from the door. (Run a sharp knife around the edges so the new paint will not be cracked and broken.) Disassemble the lock. Using a wire brush, scrape the paint from the parts. In extreme cases, you may have to resort to a small knife to do the job or else soak the individual parts in paint remover. Dry each part thoroughly. Check for rust and worn parts and replace as needed. Assemble and mount the lock.

Use paint remover only as a last resort, since it leaves a residue that attracts dust and lint. When you use paint remover, you must clean each part before assembly.

Locks that are difficult to operate usually have not been lubricated in a long time, if ever. Never use oil to lubricate a lock. The professional approach is to use a flake or powdered graphite. Apply the lubricant sparingly. Remember, a little bit goes a long way, and this is especially true of graphite. Should you overuse it, you may have to explain why there is a dark patch on the carpet that cannot be cleaned...graphite stains are almost impossible to remove.

WARDED KEYS

Warded keys are made of iron, steel, brass, and aluminum. Iron and aluminum keys have a tendency to break or bend within a relatively short time; steel and brass keys can outlast the lock. The warded key has seven parts, as shown in Fig. 4-8. The configuration of the box, length of the shank, and the relative thickness of the

Fig. 4-8. The parts of a warded key.

shoulder are not critical to the selection or the cutting of the key.

Types of Warded Keys

Warded keys come in various types: the *simple* warded key, the *standard* warded key, the *multicut* key, and the *antique* key. Simple warded keys are often factory-made precut keys which fit several different keyholes. Multicut keys, on the other hand, are designed for specific locks. The standard warded key is usually mass-produced, but it has more precut ward and end cuts than the simple warded key. Standard warded keys can be easily converted into master keys by cutting. The antique key may have several kinds of cuts: ward cuts, end cuts, and even side (or bullet) groove cuts extending the length of the bit. Antique keys usually go to older locks, but these keys are still manufactured.

Selection of Key Blanks

A key blank is a key that has not been cut or shaped to fit a specific locking mechanism. When selecting a blank for a duplicate warded key, the following should be considered:

Pin Size—The pins of both keys (original and duplicate) must be the same diameter. If your eyesight cannot correctly determine whether or not they are the same size, you can use either calipers or a paper clip. Use the calipers to compare the diameter of the original key with the diameter of the duplicate. You can use a paper clip that has been wrapped around the original key to check the diameter of the duplicate.

Length—From the collar to the end of the pin, both keys should be approximately the same length. This is important because this portion of the key enters the keyhole and operates the lock.

Height—The height of the bitting (cuts in the bit) must be the same on both keys. If the bitting in the duplicate is higher, it should be filed down; if it is lower, another blank should be selected.

Bow—The bows need not be identical, but generally should be closely matched.

Width—The width of the bittings should be approximately the same. If the bitting on the duplicate is too wide, the extra thickness may prevent the duplicate from entering the lock.

Thickness—The thickness of the bits should be the same. If the original key bit is tapered the bit of the duplicate should also be tapered. You may have to select a blank with a thick bit that you can file down to the correct taper.

Table 4-1. Standard Wire Gauges (inches and millimeters). 1 mm = 0.03937″; 1″ = 25.4 mm.

Standard Wire Gauge Number	Inches	Millimeters
4/0	0.400	10.16
3/0	0.372	9.45
2/0	0.348	8.84
0	0.324	8.23
1	0.300	7.60
2	0.276	7.01
2½	0.264	6.71
3	0.252	6.40
3½	0.242	6.15
4	0.232	5.89
4½	0.222	5.64
5	0.212	5.38
5½	0.202	5.13
6	0.192	4.88
6½	0.184	4.67
7	0.176	4.47
7½	0.168	4.27
8	0.160	4.06
9	0.144	3.66
10	0.128	3.25
11	0.116	2.95
12	0.104	2.64
13	0.092	2.34
14	0.080	2.03
15	0.072	1.83
16	0.064	1.63
17	0.056	1.42
18	0.048	1.22
19	0.040	1.02
20	0.036	0.91

If you don't have a micrometer, Table 4-1 showing standard wire diameters can help you determine the approximate diameter of warded key blanks. You can take a standard piece of wire with a known diameter and compare it with any key blank. You can also use a drill to determine the approximate diameter of warded key blanks. Insert the pin of a blank into the hole that matches the blank's diameter.

Duplicating a Warded Key by Hand

1. Select the proper key blank.
2. Wrap a strip of aluminum, approximately 1½″ ×2¼″ or 2½″ about the pin and bit of the original key with one edge against the collar (Fig. 4-9).
3. Clamp the original key (wrapped in the alu-

minum) into a vise, bitting edge up. Ensure that the aluminum fits snugly about the key bit and pin.

4. Cut off excess aluminum around the bit of the key. Remove the aluminum strip and smoke the key with a candle.

5. After properly smoking the key, place the strip back on the bit and reclamp.

6. Using a warding file, cut down the aluminum strip—*not the original key*—until it is in the shape of the original key. Since the aluminum is easily bent, use the file only in one direction—away from you. The stroke should be firm and steady at first. As you file closer to the cuts of the original key, the strokes should be shorter and lighter. When the file just barely touches the original key and starts to remove the candle black, *stop and go no further*. If the candle black is removed and the shiny surface of the original key is revealed, you know that you have filed too deep.

7. Fit the aluminum strip onto the key blank.

8. File down the exposed areas on the bit until it matches the outline of the aluminum strip. Be careful not to cut into the strip. Again, use shorter and lighter strokes as you get closer to finishing each cut.

9. If the original key has a side groove that matches a keyhole ward, this too must be cut. Use another strip of aluminum. If the original key has two grooves you must wrap the strip around the bit so that both grooves are covered.

10. Using a scriber, scratch the metal strip to indicate the top and bottom of the groove(s). Fit the strip onto the key blank and mark the positions of the groove(s) on both ends of the bit. By connecting the marks with lines, you know exactly where to file.

11. To determine the depth of a groove cut, put one edge of a metal strip into the groove of the original key and scribe the depth of the groove on it. This mark will be

Fig. 4-9. File into the aluminum strip in the direction shown.

your depth guide when filing the groove on the duplicate key.

The Impression Method

The term *impression* implies a duplicate key that is made from the lock itself as opposed to copying the original key. Impressioning is a skill that sets the locksmith apart from the individual who merely duplicates keys in the local five-and-dime store. It is a skill that is an extremely important and valuable asset.

Impressioning is the ability to decipher small marks made on a smoked key blank that has been inserted into a lock and turned. Interpretation of the marks tells the locksmith what cuts to make, where to make them, and how deep they must be. The advantage of impressioning is that you need not disassemble a lock or remove it from a door to make a key.

The first cut allows that blank to enter the keyhole. Refer to the previous section for information on what is required to properly make such a cut. Since you do not have the original key, smoke the end edge of the blank and insert it into the keyhole so the edge comes into contact with the case ward. Scribe mark the top and bottom of the ward on the blank. Remove the key; the candle black that was removed indicates the depth of the cut and the scribed marks show the position of the cut on the blank. Transfer the scribe marks to the near end of the bit and draw lines connecting the two pairs of marks. As in the previous section, use a small piece of metal to make a depth gauge so you will not file too deeply. Mount the blank in the vise and cut the ward slot. When you're finished, the key should pass the case ward.

The next step is the preparation of the blank for impressioning the internal wards. Recall that the key must pass certain side wards. When the lock is assembled, how can you be sure exactly where to cut the key so it will pass these wards? You can't. Thus, you will have to prepare the key blank. Two methods may be used; the first is to coat the key with a thin layer of wax. This is unprofessional and can harm the lock; the wax may clog the mechanism and require that the lock be disassembled and cleaned *at your expense*—you can't charge a customer for your mistake. The professional method is to smoke the key. The smoking must be thick enough to form a stable marking surface. If, for example, the blank is thicker than it should be, enough blackening must be present to give true readings. A thin coat will speckle as you turn the key and send you on wild goose chases.

The technique for impressioning the key is:

1. Insert the key into the lock and turn it with authority.

2. Remove the key. You will notice one or more bright marks where the blackening has been removed. These marks indicate obstructions and you will have to file the appropriate cut for the key to pass.

3. Blacken, insert, turn, and remove the key; if a mark is present below the point you filed, the cut is too shallow. Carefully file it deeper.

4. When the cut is deep enough, use emery paper and clean off the fine burrs left by the file. This is the professional way.

5. Reinsert the key and turn it again. It should pass the wards. If it tends to stick slightly, a quick pass over the cut lightly with the warding file will alleviate this problem. Note: The cut should be square on all sides. Think of each ward cut as a miniature square, perfect and even on all sides.

6. Smoke the key, insert it, and turn it. The edges of the cuts should be shiny. The brighter the spot, the greater the pressure at that point. As the individual cuts are filed deeper, the bright spots grow dimmer. As this happens you know that you are close to the point where you should stop filing.

Each time you insert the key to test the depth of the cut, you must be more cautious. You are working "in the blind;" a small overcut means the depth is permanently wrong on the blank and you will have to start over. This is the reason for making a few light, but firm, strokes as you near the completion of each cut. Take your time to make one perfect key instead of rushing the job and making many incorrectly cut keys.

The cuts on the key should be as deep and as wide as necessary, and no more. Overly large cuts interfere with the action of the lock, and may force you to cut another blank.

Once the key is completed, it must be *dressed out*. I mentioned earlier using emery paper to remove burrs before they break off and fall into the lock. Now polish the entire key with emery paper. You might also give it a light buffing on your wheel for appearance. It does you no good to give the customer a dirty and smudged key. Show pride in your work!

Pass Keys

Skeleton-type pass keys (master keys) are sold in variety stores. These keys will fit many old locks and more than a few new ones. As such, they are a convenience. But no reputable locksmith stocks them. Why? The ethics of the profession forbid it. You certainly do not want to supply someone with a key that could open his neighbor's lock. Nor should you duplicate a pass key without authorization from the owner. Be leery of a customer who wants a key duplicated, but with additional cuts. Locksmiths have lost their licenses for less. Don't let it happen to you.

Chapter 5

The Lever Lock

Lever locks have many uses in light security roles. Available in a variety of sizes and shapes, these locks are found on desks, mailboxes, lockers, bank deposit boxes, and so on.

Figure 5-1 illustrates a popular example. The circular "window" on the back of the case is a locksmith's aid—it reveals the heights of the levers without the necessity of dismantling the lock. Thanks to the window, it is relatively easy to make a key "in the blind."

PARTS

A lever lock has six basic parts:

☐ Cover boss.
☐ Cover.
☐ Trunnion.
☐ Lever tumblers (top, middle, and bottom).
☐ Bolt (bolt stop, notch, and post).
☐ Base.

These parts are shown in Fig. 5-2.

OPERATION

This lock requires a standard flat key. When the key is turned, the various key cuts raise corresponding lever tumblers to the correct height. As the levers are raised, the gates (Fig. 5-3) of the levers align and release the bolt. The bolt stop (some call it the post) must pass through the gating from the rear to the front "trap" or vice versa, either unlocking or locking the lock.

LEVER TUMBLER

The number of lever tumblers varies. Most locks have no more than five, although deposit box locks have more. The lever tumbler consists of six parts:

☐ Saddle (or belly).
☐ Pivot hole.
☐ Spring.
☐ Gating slot.
☐ Front trap.
☐ Rear trap.

Each lever is a flat plate and is held in place by a pivot pin and a flat spring. Each lever has a gate cut into it. The gates are located at various heights either with the saddle aligned for all levers or staggered. The latter approach has long since been antiquated. When the levers are raised to the proper position, the gates are open and the

Fig. 5-1. Note the "window" on the cover of this lever lock.

Fig. 5-2. A lever lock in exploded view. This particular lock has three tumblers; others may have a dozen or more.

bolt post can be shifted from one trap to another, thus locking or unlocking the lock.

Since the bolt post meets no resistance at the gating, the lock will work properly. On some designs the edge of the lever has serrated notches. The bolt post has corresponding notches. The notches on the lever and the bolt jam together if an improperly cut key is used. This effectively stops the bolt from passing through the gating and keeps the lock secure. Only a perfectly and properly cut key will open this type of lock. This feature adds immensely to the security of the lock.

Manufacturers have, over the years, developed a variety of lever types (Fig. 5-4). The operating principle is the same for all of them.

The width of the gate is also a critical factor in the operation of the lock. Some gates are just wide enough for the pin to pass through. A duplicate key, even slightly off on a single cut, will not work on the lock.

The saddle of the lever is also important. Recall from Chapter 1 that staggered saddles make it possible to cut a key by observation. In the case of modern lever tumblers, the gate traps have different heights, leaving the saddles in perfect alignment.

Gating changes are made by two methods. The most common is to substitute a lever tumbler with a different gate dimension. Locksmith supply houses stock a variety of levers, so all you are required to do is change the original for one with a higher or lower gate. The second method is to alter the tumbler by filing the saddle. This approach is used with levers whose movement is re-

Fig. 5-3. Lever-tumbler nomenclature. The key bears against the saddle.

Fig. 5-4. Lever-tumbler styles. The gating may be open or closed (as in the lower right example).

stricted at the gate. Unless the tumbler gating varies greatly, the curvature of the saddle must vary with the shape of the key.

When working on the lever lock, as on other types, it is best to have two available: one for disassembly to determine internal working parts, and the second for actual problem-solving use.

As I mentioned earlier, the typical lever lock contains two, three, five, or possible six levers. Bank deposit box locks may have as many as 14. Lever locks can be keyed individually, alike (two or more with the same key), or master keyed, depending upon the wishes of the buyer.

DISASSEMBLY

General-purpose lever locks come in three styles. In order of popularity these are:

☐ Solid case, usually spot-welded or riveted.
☐ Pressed form with the back and sides of the case one piece—small tabs from the sides bend to hold the cover in place.
☐ Cover plate secured by a screw.

Riveted or spot-welded locks should be discarded when they fail. It is cheaper to purchase a new lock; the time required to drill out the rivets or chisel through the spot welds costs the customer more than the lock is worth.

Locks secured by tabs can be disassembled and reassembled quite easily. To disassemble, insert a thin tool or small screwdriver under the flanges and pry upward. Remove the cover. Look for a small object jammed in the keyway; if something is found, remove it. At times like this you may wonder if being a locksmith is really worth it, but a locksmith must have patience with the small jobs as well as with the big ones.

Another common problem with lever locks is a broken spring. If it is not the top lever, then carefully remove each lever, in turn, placing them in a logical order so that they may be assembled as they were found. Remove the lever whose spring is broken. Select a piece of spring steel from your inventory. Cut and bend it to shape. If you have purchased an assortment of ready-made springs, select the proper one and replace the broken spring with a new one. Replace the levers and reassemble the lock.

SAFE DEPOSIT BOX LOCKS

Lever safe deposit box locks normally have a minimum of six and upwards of twelve or even fourteen levers. Two keys are required. One key is assigned to the individual who rents the box; the second or guard key is held by the bank. Both keys are needed to open the lock.

Disassembly is simple. Remove the screws and lift off the plate. You will notice that the lock is constructed differently than previously discussed lever locks. There are two sets of lever tumblers; two bolt pins must pass through the lever gates at the same time. Note the unique lever shape (Fig. 5-5).

Many safe deposit locks have a compression spring bearing against the upper lever that forces the lever stack down, allowing no play between them. This spring is an integral part of the lock's security mechanism. Without the spring, the levers would be able to move a fraction of an inch when a key or pick "irritates" them. This movement is enough to provide clues for the lockpick artist.

This lock, as stated earlier, requires two keys to open it. If one key could be turned to the *open* position by

Fig. 5-5. Notched tumblers are a security feature in some locks.

itself, it would mean that the key was faulty, the lock mechanism was in some way evaded, or the bolt post was bent or broken.

The levers have a V-shaped ridge that matches a similar V cut in the bolt post. Another key, even one that is 0.001″ off on any cut, will mesh the notches in the post and lever. This is another built-in security feature of these locks.

Broken springs are the main difficulty. Or the levers may be at fault. The saddles may wear enough to affect the gate. In cases such as this, replace the lever with a new one. Do not file the gate cut wider to compensate. Sometimes a pivot post will work loose. Repair by rapping the post with a light hammer or by brazing. Never use solder. Slightly bent parts can be straightened out, but it is best to replace the entire damaged part.

Safe deposit locks are serviced by locksmiths who enjoy high standings within the community. Unless an emergency arises, a beginning locksmith or one who is a newcomer to the area would not be given the job. But this reluctance is only a matter of the conservatism of bankers. It does not reflect upon the skills of students and those working locksmiths who have had only a few months experience. Many of these men and women could repair a safe deposit box lock without difficulty.

SUITCASE LOCKS

Suitcase locks are essentially simple but have been built to accept a staggering variety of keys. Ninety-nine locks out of a hundred are warded, with a primitive bolt mechanism to keep the case closed. Only a few, such as the Yale luggage lock, use a lever-type mechanism. These locks offer better security than warded locks.

The lock size, the depth the key is inserted, and the number of cuts are important clues to the lock type. A key for a lever lock will go ¼″ or deeper into the lock before turning. Warded locks are shallower.

Opening suicase locks is relatively simple. Many times one key will open several different suitcase and luggage locks. Many of these locks on the market are not designed for security. By cutting down almost any suitcase key, it is possible to make a skeleton key for emergencies.

LOCKER AND CABINET LOCKS

The Lock Corporation of America has come up with a new locking spring design accompanied by a free-turning keyway that combine to make the locks extremely jam-proof and pick resistant in the flat key type of locks. Figure 5-6 shows such a lock.

These locks are made to last (and also carry an unconditional guarantee). They are constructed of a one-piece heavy duty steel case that will stand up to years of use, even under the most corrosive conditions. The hard wrought steel keys and the lock case have been electroplated and dichromated to ensure numerous years of trouble-free service.

Another advantage to these locks are that they have been designed to fit into current master key systems. The lock will fit left or right-hand steel locker or cabinet doors that have the standard piercings.

The keys can be removed from either the deadbolt or

Fig. 5-6. Flat key deadbolt locker lock. (courtesy Lock Corporation of America)

Fig. 5-7. Flat key lever tumbler locker lock. (courtesy Lock Corporation of America)

The model 5001 MK locker lock is a built-in flat key lock with a deadbolt locking mechanism; the key is removable in either the locked or unlocked position. A free-turning plug (with a minimum of five lever tumblers) provides for a minimum of 1400 different key changes. The locks (within a key change range specified) can be master keyed if desired. The lock will not be operated by keys of another lock within the specified key range, except a master key. The casing is of wrought steel and is 1⅜″ by 1⅝″ with top and bottom attaching ears to fit the standard steel locker piercings. All parts are of phosphor bronze, zinc or rustproofed steel. The backset is 1″ or less, and the solid locking bolt is ¾″ by ¼″ with a 5/16″ throw.

Specifications are the same for the 5000 MK except for a key removable in the locked position only; the 4000 MK, differs only in having a solid spring bolt locking mechanism. The internal parts' relationship and positioning are illustrated in Fig. 5-9.

Looking at the inner workings of the LCA torsion lever flat key lock, we find that it contains a patented multi-movable lever flat key operated mechanism that have the latest improvements to the world's oldest and still most popular means of security—the warded lock.

Seven levers (including one master lever, if required) engage and restrain the locking bolt. The novel uni-spring construction is arranged in two banks of seven staggered interlocking arms constantly exerting a force on all the levers. By inserting and turning the precision cut key in the rotatable key guide, the latched levers thereby are moved from securing the deadbolt, enabling the lock to be opened.

Of note is that the model 4000 (basically the same) has a bolt split into two parts, including a separate spring which allows the bolt to retract without using a key to close or slam shut.

The grooved key built-in torsion tumbler cylinder lock (Fig. 5-10) has the following specifications:

☐ Standard three-hole piercings for mounting.
☐ Beveled locking bolt, moving horizontally.
☐ Locking bolt is ¾″×⅜″ with a 5/16″ throw.
☐ Over one million computerized key changes.
☐ Eight nickel silver torsion tumblers (to provide the maximum security).
☐ Regular, sub-master, master keyed, keyed different, keyed alike or group keyed.
☐ The master key design prevents the use of a regular key as a master key.
☐ The keys are embossed, nickel plated, and grooved.

springbolt locks when the locker door is opened or closed. The deadbolt locks are available in models with key removable in the locked or unlocked mode, or with the key removable in the locked position only. Also, with these locks, you have a self-locking springbolt with the key removable in the locked position only.

The series 4000 and 5000 flat key built-in lever tumbler (Fig. 5-7) specifications are as follows:

☐ Mounts in a standard three-hole piercing on steel lockers, cabinets and other shop equipment.
☐ Solid locking bolt moves horizontally with a 5/16″ throw.
☐ Master keyed, keyed different, keyed alike, or group keyed.
☐ Key removable when locked, or when locked or unlocked (must be specified when ordered).
☐ .062″ gauge heavy-duty one-piece steel case.
☐ Free-turning keyway.
☐ Phosphor bronze locking springs.
☐ Five hard wrought steel levers.

The dimensions for the lock and the standard locker door punchings are shown in Fig. 5-8.

Fig. 5-8. Specifications for the LCA series 4000 lever tumbler flat key lock. (courtesy Lock Corporation of America)

Fig. 5-9. Internal parts positions for the 4000 and 5000 series flat key lever locker lock. (courtesy Lock Corporation of America)

Fig. 5-10. Built-in torsion tumbler cylinder lock. (courtesy Lock Corporation of America)

Fig. 5-11. Series 6000 torsion tumbler lock with inset illustrating the snap-on face plate. (courtesy Lock Corporation of America)

"SNAP-ON" INSTALLATION

3/8 × 50°

1/4"

5/16" THROW

7/32"

FOR
7/32"
HOLE

1 23/32"

55/64"

2 3/16"

3/4"

3/8"

1 3/32"

45/64"

STANDARD LOCKER
DOOR PUNCHINGS

7/32"

21/32"

2 3/16"

1 3/32"

Fig. 5-12. Specifications for series 6000/7000 "snap-on" installation lock (courtesy Lock Corporation of America)

Fig. 5-13. Eight torsion tumbler cylinder drawer lock. (courtesy Lock Corporation of America)

☐ Cylinder plug of solid brass.
☐ .062″ gauge heavy-duty one-piece steel case construction.

The 6000 and 7000 series have a patented snap-on face plate for installation (Fig. 5-11). Technical drawings for the snap-on installation are at Fig. 5-12, along with the locker door hole punchings. Note that the series 6000 has a spring bolt, while the 7000 operates with a deadbolt.

The 7001 locker lock has a built-in high security type of ruggedness with a deadbolt locking mechanism and the cover plate; it is master keyed with the key removable in either the locked or unlocked position. Over 100,000 possible different key changes are available. The bolt is ¾″ by ⅜″ with a 5/16″ throw.

The LCA drawer and cabinet lock (Fig. 5-13) is constructed of high-pressure solid Zamak unicast casing and base and has a ⅞″ diameter cylinder case for use in up to 1⅜″ thick material.

The torsion tumbler is based on a new and different design concept. Each of the eight nickel silver torsion tumblers is manufactured with its own integral spring; it is made strong and durable by an exclusive forging process. This tumbler-spring design will provide a much greater protection against unfavorable climatic condi-

tions such as temperature and humidity changes because the torsion spring is heavier and sturdier than the finely coiled springs required by either pin or disc tumbler cylinders.

Variation in the eight tumblers (arranged four to each side of the cylinder) and in the double bitted key provide over one million key change possibilities. The unique open construction of a one-piece tumbler and spring eliminates clogging due to accumulated dirt and small particles, and offers long and high performance reliability, smooth operation, and improved pick resistance.

Better master key control and protection are provided by an entirely different master key configuration. This special key design prevents the fashioning of a master key from regular keys or blanks, thus reducing the possibility of unauthorized entry. Practically unlimited master key changes are available without diminishing the security of the cylinder.

Figure 5-14 provides details of the maximum security eight torsion tumbler cylinder in an exploded view. Figure 5-15 shows the critical dimensions of the cylinder.

The LCA push lock for sliding door and showcases uses the same patented torsion tumblers and solid brass cylinder plug, and has a stainless steel cap (Fig. 5-16). Three hundred key changes are standard, but unlimited changes are available.

The various series 3300 and 3400 locks can have a push bolt, automatically locked by pushing the cylinder inward or releasing the cylinder and catch by turning the key 90°. The bolt is ⅜″ in diameter with a 15/32″ throw and a ⅞″ cylinder case for use in up to 1⅛″ thick material mountings.

The push-turn bayonet type lock in this LCA series has a different strike plate (Fig. 5-17). It is locked by turning the key 90°, pushing the cylinder in and rotating the key again to lock the "T" type plunger behind the strike plate, thus preventing the doors from being pried apart. The cylinder locks in either the in or out positions for further protection. With the 3350 model lock, a 7/16″ diameter case hardened steel covered bolt is available. Figure 5-18 shows the 3300 model technical dimensions, while Fig. 5-19 provides the same information on the 3400 model.

For the locksmith concerned with businesses, these locks are a stock in trade, constantly proving themselves over and over.

LEVER LOCK KEYS

Unlike the warded or cylinder key, the lever lock

Fig. 5-14. Exploded view of the maximum security 8-torsion tumbler cylinder. (courtesy Lock Corporation of America)

key used by the average person almost never contains a keyway groove running along its side; it is a flat key. A number of different flat keys exist. Figure 5-20 and 5-21 show some of them. Figure 5-22 identifies the various parts of a typically cut lever lock key.

The lever lock key can be cut by hand or machine. In order to make the key you must have the proper key blank. The three critical dimensions of the lever lock key are the thickness, length, and height. If the key blank is slightly higher or wider than the original, the blank should be filed down to the proper size. If it is thicker, select another blank. Filing down the thickness of the blank weakens it structurally.

Laying the keys (original and blank) side by side, run your finger lightly across them. If the blank is thicker than the original, you will feel your finger catch as it passes from one to the other. When you insert the key into the keyway, it should not be tight fitting or bind.

The first cut that must be made is the *throat cut*, which enables the key to turn within the keyway. To do this, insert the blank in the keyway and scribe each side of the blank where it comes in contact with the cover boss. Determine the point where the trunnion of the lock turns. Draw a vertical line there to indicate the depth of the throat cut. Remove the key from the lock, place it in a vise, and use a 4″ warding file to cut alongside the vertical

Fig. 5-15. Critical cylinder dimensions of the 8-torsion tumbler cylinder lock. (courtesy Lock Corporation of America)

Fig. 5-16. Sliding door and show case locks. (courtesy Lock Corporation of America)

Fig. 5-17. Grooved key push lock for sliding doors and display /show cases. (courtesy Lock Corporation of America)

line to the proper depth. Cut on the side of the line towards the tip of the key.

There is a small round window in the back of most lever lock cases. The window is positioned so you can see where the bolt pin meets the lever gates. You can get a general idea of the proper cuts to be made on a key blank by observing the lever action through the window.

After the throat cut is filed, the other cuts must be made. To do this, follow these steps:

1. Smoke the blank, insert it in the keyway, and turn. Remove it. The lever locations will be marked on the blank.

2. File the marks slightly, starting with the marks closest to the throat cut.

3. Insert the key and turn it. Notice (through the window) the height that each lever comes up and the position of the pin in relation to the lever gates. The distance each gate is from the pin indicates the depth the key should be cut for each particular lever. File each key cut a little at a time, periodically inserting and turning the key to check the gate/pin relationship. Be sure to file *thin* cuts. Continue until you can insert the key and have the gate and pin line up exactly.

Fig. 5-18. Technical dimensions for the sliding door push button lock. (courtesy Lock Corporation of America)

LCA MODEL #3400

LOCKED POSITION

AT UNLOCKED POSITION

AT LOCKED POSITION

7/8″

TRAVEL 15/32

TURN KEY 90° TO UNLOCK

3/32″

1⅛″

½″

2″ DIA.

7/8″ DIA.

DIA

½″

¾″

1½″

STRIKE PLATE

DRILLED & CTSK FOR #6 FLAT HEAD SCREW

Fig. 5-19. Technical dimensions for the push and turn showcase lock. (courtesy Lock Corporation of America)

4. If the key in the keyway binds, observe the levers. One or more may not be correctly aligned. If the pin is too high, you have cut too deeply; if the pin is too low, you have not cut deeply enough. It may be necessary to resmoke the blank and reinsert it. The point where the key binds hardest will be indicated by the shiniest spot on the key. Just a touch with the file will usually alleviate the problem. Insure that each filed cut is directly under its own lever.

At this point, you should have corrected any variations between the original key and the blank. The dimensions must be identical: the key height, thickness, and the length.

Fig. 5-21. Taylor Lock Company supplies these and other lever-lock blanks.

Fig. 5-20. Typical lever-lock keys.

Fig. 5-22. Lever-lock key nomenclature.

You will need a vise for holding the two keys, a small C-clamp, a warding file, a candle, and a pair of pliers. Follow this procedure.

1. Holding the original key over a candle flame, smoke it thoroughly.

2. Allow a few minutes for the key to cool and clamp the original and the blank at the bows. Most locksmiths use a C-clamp for this initial alignment.

3. Once aligned, place the key and blank in a vise. If you wish, you can leave the C-clamp in position.

4. Using the warding file, make the tip cut first. File in even and steady strokes, bearing down in the forward, or cutting stroke. Keep a careful eye on the original. Stop when the file just disturbs the blackening.

5. Once the tip cut is completed, move to the next cut.

6. Remove the keys from the vise and inspect the cuts. Each should be rectangular and flat.

7. Using emery paper, lightly sand away the burrs on the edges of the cuts. Wipe off the candle black from the original key.

8. Test the duplicate in the lock. Should it stick, blacken the duplicate and try it again in the lock. Breaks in the blackening will show where the key is binding. A light stroke with the file should correct this.

Chapter 6

The Disc-Tumbler Lock

The *disc-tumbler* (or *cam*) lock gets its name from the shape of the tumblers. These are about as secure as lever locks, and as such are superior to warded and other simple locks. However, disc-tumbler locks are far less secure than pin tumbler locks.

Disc-tumbler locks are found in automobiles, desks, and in a variety of coin-operated machines. Some padlocks are built on this principle. Because the cost of manufacture is very low, replacement is cheaper than repair.

While similar to the pin-tumbler lock in outside appearance and in the broad principle of operation, the internal design is unique.

The disc tumblers are steel stampings, arranged in slots in the cylinder core. Figure 6-1 shows a typical disc tumbler. The rectangular hold, or cutout, in the center of the disc matches a notch on the key bit. The protrusion on the side, known as the *hook*, locates the spring. The disc stack is arranged with alternating hooks, one on the right of the cylinder, one on the left. The complete lock is illustrated in Fig. 6-2.

OPERATION

The disc lock employs a rotating core, as does the more familiar pin-tumbler design. The disc tumbler core is cast so that the tumblers protrude through the core and into slots on the inner diameter of the cylinder. The core is locked to the cylinder so long as the tumblers are in place.

The key has cuts that align with the cutouts in each tumbler. The key should raise the tumblers high enough to clear the lower cylinder slot, but not so high as to enter the upper cylinder slot. In other words, the correct key will arrange the tumblers along the upper and lower shear lines (Fig. 6-2). The plug is free to rotate and, in the process, throw the bolt.

The key resembles a cylinder pin-tumbler key, except that it is generally smaller and always has five cuts. A cylinder pin-tumbler key may have six or seven cuts.

DISASSEMBLY

Good quality disc locks feature a small hole on the face of the plug, usually just to the right of the keyhole. Insert a length of piano wire into the hole and press the retainer clip. Turn the plug slightly to release. The key gives enough purchase to withdraw the plug. If a key is not available, you can extract the plug with the help of a second length of piano wire inserted into the keyhole. Bend the end of the wire into a small hook. Other locks attach the plug to the cylinder with the same screw that

Fig. 6-1. Disc-tumbler lock in cross section. The position of the cutout determines the depth of the key cut.

secures the bolt-actuating cam. Others (fortunately, a minority) have the plug and cylinder brazed together. File off the brass.

KEYING

Manufacturers have agreed upon five possible positions for the cutouts relative to the tumblers. Keying is a matter of arranging the tumblers in a sequence that matches the key cuts. Once the sequence has been identified, install the tumbler springs over their respective hooks and mount the tumblers in the plug. The tumblers are spring-loaded and, until the plug is installed in the cylinder, are free to pop out. Lightly stake them in place with a punch or the corner of a small screwdriver blade. One or two pips are enough since the tumblers will have to be broken free once the assembly is inside the cylinder. Inserting the key is enough to release the tumblers.

SECURITY

These locks have no more than five tumblers and each tumbler cutout has five possible positions. These variations allow 3125 or (5^5) key changes, at least in theory. In practice the manufacturer will discard some combinations as inappropriate and may further simplify matters by limiting the key changes to 500 or less. Obviously, disc-tumbler locks are not high-security devices.

CAM LOCKS

Cam locks are used for many general and special purposes for a variety of needs in business (Fig. 6-3). More than likely, you will see approximately 90 percent of all these units in office situations. From the face (front) these locks look pretty much the same after they have been installed. Beyond that point, however, they may differ in various regards. The illustrations on the next several pages provide exploded views of different models and the various parts associated with each model. These models are all five-disc tumbler locks, each with a possibility of 200 different key combinations. Because of the key variations possible, these locks are often master keyed prior to purchase by the customer. Only after disassembly or by viewing the tumblers through the keyway will you know for sure whether or not the lock is master keyed (Figs. 6-4 through 6-9.)

Two other core type cam locks are shown in Figs. 6-10 and 6-11. These cam locks have a removable core, and are also of the seven-pin variety which uses the Ace-type circular key.

The variety of cams is used with these locks. Figure 6-12 is a chart showing the various types of cams available. From the types available you can find a cam for the

Fig. 6-2. A disc-tumbler lock in exploded view.

Fig. 6-3. Five-tumbler wooden desk cam lock. (courtesy Dominion Lock Co.)

lock that will meet every possible need. Included are the length and offset specifications also.

Figure 6-13 shows the hook, bent, double-ended and other miscellaneous cams that you may come across in the course of your work.

The standard thickness of the cam is 3/32″ (2.667 mm); they are of steel, with many being cadmium plated for durability and longevity.

READING DISC AND LEVER LOCKS

It's not unusual for a locksmith to be asked to make a key for a lock when the original key has been lost or misplaced. If the lock does not have a code number on its face or if the owner neglected to write down the number on the key, the locksmith has three choices. He can pick the lock, impression a key, or "read" the lock.

Like impressioning, reading a lock is a skill that must be developed through patient practice. You cannot expect to master this skill quickly, nor can you expect to remain proficient in it without constant practice. At first, practice daily, then weekly or twice-weekly to maintain your skill.

When called upon to fit a key, either by impressioning or reading of the lock, the locksmith invariably looks into the lock keyway. A quick glance determines whether it is a lever, disc, or pin-tumbler lock.

Disc Locks

The view down the keyway of a disc-tumbler lock will show a row of discs with their centers cut away and staggered. The cutaways are at different heights; the discs themselves are the same diameter. The cutaway

Parts
Desk Lock

1. Nut
2. Bolt
3. Shell
4. Retainer clip
5. Spring
6. Plug
7. Springs
8. Disc tumblers
9. Key

27/32"
21.43 mm

19/32"
15.08 mm

3/4"
19.05 mm

13/16"
20.64 mm

11/32"
8.73 mm

Fig. 6-4. Examples of the common use desk locks in exploded view. (courtesy Dominion Lock Co.)

Parts
Cam Locks

1. Nut
2. Lock washer
3. Cam
4. Stop cam
5. Nut
6. Shell
7. Plug
8. Springs
9. Tumblers
10. Key

Fig. 6-5. Cam lock, exploded view. (courtesy Dominion Lock Co.)

looks like a small staircase with a surrounding wall (formed by the vertical edges) about the steps. Since each disc is the same size, only the height of the "steps" varies, and this variation is predetermined. A No. 1 disc has its cutaway toward the top of the tumbler; a No. 5 has its cutaway situated low on the tumbler (Fig. 6-14).

Here is where the skill of reading a disc lock is learned. Study the discs through the keyway. Notice the relationship of the discs to each other and to the keyhole. Through constant study and practice, which includes mixing the various discs, you will be able to determine which disc is which (Fig. 6-15).

Lift each disc and compare it to the disc in front of or behind it. To do this, you will need a reading tool. This is nothing more than a stiff length of wire about three inches long, mounted in a small dowel handle. Think of the tool as a long hairpin attached to a short piece of wood for convenience in holding. The wire should be bent slightly so that you can see the tumblers.

Insert the tool into the lock, holding it so you can see the interior and observe the discs. By shifting the tool

about, raising and lowering each disc, you can see the relationship of each disc cutaway to the next disc and to the keyway. Using your knowledge about the cutaway relationships, you will be able to decode the tumblers and cut a key for the lock.

You might ask yourself, "How do I know where to cut the key—and how deep?" Recall that when impressioning a key, you blackened the key and determined the cutting point by the pressure of the levers upon the key as you tried to turn it. The same technique applies here. Insert a blackened key and give it a slight turn. This brings the key in contact with the sides of each tumbler cutout. The cutout will leave a mark on the blank, indicating the portion of each cut.

Determining the depth of the cuts requires experience. You have already learned that cutting a key is slow, patient work. As you get closer to the proper depth, the file is moved with less pressure than before. The same care and precision is needed here. A disc with a No. 1 cut hole requires a deep key cut, as compared to a No. 4 or 5 disc, which requires a very shallow cut.

**Parts
Drawer Lock**

1. Mounting screws
2. Bolt
3. Shell
4. Retainer clip
5. Spring
6. Plug
7. Springs
8. Disc tumblers
9. Key

1 5/32"
29.37 mm

1/8"
3.18 mm

7/8"
22.23 mm

2 1/16"
52.39 mm

1 7/16"
36.51 mm

5/8"
15.88 mm

11/32"
8.73 mm

DIA.
7/8"
22.23 mm

1 7/16"
36.51 mm

Fig. 6-6. Drawer lock, exploded view. (courtesy Dominion Lock Co.)

Parts
Drawer Lock
With Removable Backplate

1. Mounting screws
2. Removable backplate
3. Bolt
4. Shell
5. Retainer clip
6. Spring
7. Plug
8. Springs
9. Disc tumblers
10. Key

Fig. 6-7. Drawer lock, exploded view. (courtesy Dominion Lock Co.)

Parts
Sliding Door Lock

1. Mounting screws
2. Case
3. Screw
4. Spring
5. Shell
6. Spring
7. Retainer clip
8. Spring
9. Cam
10. Plug
11. Springs
12. Disc tumblers
13. Key

Fig. 6-8. Sliding door lock, exploded view. (courtesy Dominion Lock Co.)

Parts
Sentry Series Cam Lock

1. Screw
2. Lock washer
3. Cam
4. Stop cam
5. Nut
6. Shell
7. Disc tumblers
8. Springs
9. Plug
10. Key

Fig. 6-9. Sentry series cam lock, exploded view. (courtesy Dominion Lock Co.)

Parts
Cam Locks

1. Nut
2. Lock washer
3. Cam
4. Stop cam
5. Cylinder
6. Springs
7. Top pins
8. Bottom pins
9. Spindle assembly
10. Nut
11. Shell pin
12. Shell
13. Key

Fig. 6-10. Ace type key cam lock, exploded view. (courtesy Dominion Lock Co.)

Parts
Removable Core Lock

1. Cam
2. Sleeve
3. Stop cam
4. Cylinder
5. Springs
6. Top pins
7. Bottom pins
8. Spindle assembly
9. Shell pin
10. Shell
11. Key

Fig. 6-11. Removable core lock, exploded view (courtesy Dominion Lock Co.)

Reference aids include various extra keys that you have collected. Since the depths of the cuts are standardized in the industry, you can have a key with a 13354 cut and, by observing the differences in the depths of each cut, know exactly how deep a No. 2 cut should be. You can obtain a depth key set from a locksmith supply house. This set has a different key for each depth, with the same cut in all five tumbler positions. Thus, a No. 2 key has five No. 2 cuts; a No. 3 key has five No. 3 cuts; and so on throughout the series. With the help of these keys and your trained eyesight, you can place a blank in a vise and cut the key by hand.

Without a depth key set you can use a variety of disc keys as guides. Select a key with the proper cut and align your blank to it. Another approach is to make your own set of depth keys. Making the set teaches you how deep to make each cut at any given position on the blank.

Lever Locks

As you already know, a lever-lock keyway is narrow

and the view of the tumblers is further restricted by the trunnion. While this is a handicap, it can, in part, be overcome with the help of an appropriate reading tool.

The positions of the lever saddles is one clue to reading the lock. The saddle of the lock can tell you quite a bit. The wider the saddle, the deeper the cut; the narrower the saddle, the shallower the cut. Using the reading tool, you can feel the different saddle widths to determine the cuts and develop some idea of the cuts and develop some idea of the key shape.

In order to do this properly, you must have an appreciation of the internal workings of the lock. Depth key sets can be useful when you are ready to cut the keys. Lever cuts are usually in 0.015—0.025″ increments.

Practice is required. Begin with a lock that has a window in it. If possible, obtain extra levers so you can change the keying of the lock at will.

Raise each lever with the reading tool and try to determine the proper position for each one. Remember the general rule: a wide saddle requires more movement

Fig. 6-12. Cam lengths and offsets. (courtesy Dominion Lock Co.)

LENGTH			
	A	2 3/8"	60.325 mm
	B	2 1/4"	57.150 mm
	C	2 1/8"	53.975 mm
	D	2"	50.800 mm
	E	1 7/8"	47.625 mm
	F	1 3/4"	44.450 mm
	G	1 5/8"	41.275 mm
	H	1 1/2"	38.100 mm
	I	1 3/8"	34.925 mm
	J	1 1/4"	31.750 mm
	K	1 1/8"	28.575 mm
	L	1"	25.400 mm
	M	7/8"	22.225 mm
	N	3/4"	19.050 mm

OFFSET			
	O	1/2"	12.700 mm
OUTSIDE	P	3/8"	9.525 mm
	Q	1/4"	6.350 mm
	R	3/16"	4.775 mm
	S	1/8"	3.175 mm
	T	1/16"	1.600 mm
INSIDE	U	1/16"	1.600 mm
	V	1/8"	3.175 mm
	W	3/16"	4.775 mm
	X	1/4"	6.350 mm
	Y	3/8"	9.525 mm
	Z	1/2"	12.700 mm

Fig. 6-13. Hook, offset, double-ended and miscellaneous type cams. (courtesy Dominion Lock Co.)

Fig. 6-14. Disc tumblers have five variations.

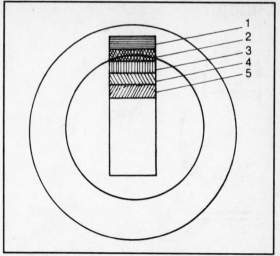

Fig. 6-15. The position of the tumblers will give you a general idea of the key profile.

than a narrow saddle. Once you are satisfied that you have read the lock, disassemble it and examine the lever tumblers. You may have misread the tumblers, but do not be disheartened. The only way to achieve competence in this skill is practice and more practice.

Other tumblers are designed with uniform saddle widths. Keying is determined by the positions of the gates in the tumblers. These locks can be quite difficult to read.

Working with the reading tool, raise one of the levers as high as it will go. While this movement is not a direct indication of the depth of the key cut, it is important. The amount of upward movement establishes the minimum key cut; a shallower cut would jam the levers against the top of the lock case. In addition, the individual gates and traps are usually in some rough alignment. You will usually find that two are in almost perfect alignment.

A cut for a lever that has the post on the upper half of the trap will be shallow; a cut for a lever that has its post on the lower half will be deep. A lever that has its gate in the intermediate position will require a key cut between these extremes.

Once you have established the general topography of the cuts, refer to your set of depth keys for the exact dimensions. Established locksmiths have reference manuals that may simplify the work.

Chapter 7

Schlage Wafer-Tumbler Locks and Keys

The wafer-tumbler design, by Schlage, offers the greatest improvement ever over the disc-tumbler lock. The wafer-tumbler cylinder is not die-cast but is made of tubular steel, manufactured to stringent tolerances for minimal slippage and the greatest exactness in alignment. Wafer tumblers are made of heavier metal than disc tumblers. This ensures long life. Wafer-tumbler locks have more tumblers than most other locks: there are eight wafers in each lock.

THE KEYWAY UNIT

The following material is furnished by the Schlage Lock Company. It covers everything a good locksmith needs to know about the Schlage wafer-tumbler lock and key.

Disassembly and Removal

1. Refer to Fig. 7-1. Depress the catch (A1) through the hole in the shank of the inside knob (A) with a screwdriver to release the inside knob from the spindle (E).

2. With the knob removed (Fig. 7-2), position the screwdriver into the small notch (B1) usually located on the bottom edge of the inside rosette (B) and, with a prying motion, snap off the inside rosette.

3. Refer to Fig. 7-3. Remove the two machine screws (C) so the inside mounting plate (D) will slip off over the inside spindle (E). The lock will now slip out of the hole.

4. Refer to Fig. 7-4. The lock housing (F) is attached to the lock by a small cotter pin (G) or by twisted lugs. Remove the cotter pin or straighten out the lugs. Lift the housing above the lugs and, with a slight turn, rotate the housing ¼" and remove it.

5. With the housing removed (Fig. 7-5), the lock frame is now exposed. Hold the lock in both hands, positioning your fingers as shown. When performing this operation, hold your palms carefully around lock to prevent springs from escaping the retractor slide. To remove the thrust plate (I), press forward with your thumbs against the frame tabs (H1) and push upward with your index fingers against the thrust plate. This will disengage plate.

6. Refer to Figs. 7-6 and 7-7. In order to free the plunger unit (L) as you remove this assembly, it is necessary to push the slide (J) all the way to the rear against the compression of the two slide springs (K) with your thumb.

7-1. Schlage wafer-tumbler lock disassembly, Step 1.

Fig. 7-4. Schlage wafer-tumbler lock disassembly, Step 4.

Fig. 7-2. Schlage wafer-tumbler lock disassembly, Step 2.

Fig. 7-5. Schlage wafer-tumbler lock disassembly, Step 5.

Fig. 7-3. Schlage wafer-tumbler lock disassembly, Step 3.

Fig. 7-6. Schlage wafer-tumbler lock disassembly, Step 6.

7. After the inside spindle (E), thrust plate (I), and plunger assembly (L) have been removed, let the slide (J) and the two slide springs (K) ease forward gradually and remove them from the lock frame (H).

8. Refer to Fig. 7-8. To remove the wafer keyway unit (N) from the lock, push in on the face of the wafer keyway unit from the outside knob (M). The unit will then slide inward, where it can be removed from the knob assembly.

9. To facilitate reassembly (Fig. 7-9), remove the outside knob (M) from the lock frame by rotating the knob ¾ of a turn while pulling out.

Types of Keyway Units

Schlage wafer keyway units are made in two distinct types, type 1 and type 2. To distinguish between these two types, look first at the master wafer column. In type 1

67

Fig. 7-7. Schlage wafer-tumbler lock disassembly, Step 7.

Fig. 7-9. Schlage wafer-tumbler lock disassembly, Step 9.

Fig. 7-8. Schlage wafer-tumbler lock disassembly, Step 8.

(Fig. 7-10) the elongated slot will be at the top. Type 2 keyway has the elongated slot of the master wafer column below the V-groove.

Types of Keys

As there are different types of keyways, there are also different types of keys (Fig. 7-11). These can be recognized by looking at the tip sideways to see which portion above or below the V-groove has been cut away. If the portion above the V-groove has been cut away, this is a type 1 key. If the portion below the V-groove has been cut away it is a type 2 key. If the tip is uncut it is a type 0 key, usually used as a master or grand master key.

Fig. 7-10. The two distinct types of Schlage wafer keywav units.

Fig. 7-11. The three different types of keys for wafer keyway units.

Wafers

To more easily recognize the three types of wafers, always arrange them so that the small protrusion is upward and the opening is to the right (Fig. 7-12). Notice that each of the three types has a definite silhouette. The master wafer (N5) has a notch cut out at the base of the protrusion just inside the spring seat. The combination wafer (N8) has a protrusion on the rounded shoulder opposite the spring seat location. The series wafer (N9) has a protrusion at the top of the wafer close to the spring seat but does not have the small notch, as does the master wafer (N5). Each of these wafers performs in a different manner, and it is most important to recognize each type before it is inserted in the keyway unit.

Fig. 7-12. Silhouettes of the three kinds of wafers.

Operation of the Keyway Unit

Figure 7-13 shows two wafer keyway units set to the same combination. A is in the relaxed position (with the key out of the keyway). B has the proper key inserted. Notice in A there are four protrusions of the wafers, one at the bottom and three at the top. The first protrusion to the right of the plunger spring (N3) is the master wafer (N5). This remains out except when it is retracted by the uncut portion of the tip of the key. The cut portion of the tip is necessary to allow full insertion of the key into the keyway. The three protrusions (N9) at the top of the keyway are the series wafers. When the key is inserted, the uncut portion opposite the protrusion acts upon the series wafers to pull them into the keyway.

The remaining four wafers (N8) in the keyway are combination wafers. Both in the relaxed position (A) and with the key inserted (B) these wafers lie within the confines of the keyway unit. Therefore, cuts on the key adjacent to their protrusion are required to prevent them from being pushed out into the locking position. An improper key will fail to draw back all the protrusions of the master and series wafers and will extend some or all of the protrusions on the combination wafers.

Description of the Wafer Keyway Unit

In working with the wafer keyway unit (Fig. 7-14), hold it in your left hand with the V-grooved dividing strip (N1) facing you. In this position the spring comb is on the

Fig. 7-13. Two wafer keyway units set to the same combination, one (A) in the relaxed position, the other (B) with the proper key inserted.

underside and cannot be seen. The protrusion at the extreme left is the keyway cam (N2). To the right is the plunger spring (N3). Next to this is the keyway frame (N4), which includes the entire steel area from the plunger spring to the finished cap of the keyway unit. In this steel framework are the wafers which are activated by the insertion of the key.

The first column (N5) to the right of the plunger spring is the location of the master wafer. Notice that the slots in this column have a different proportion than do the other columns in the keyway frame. In the other columns the top slot is shorter than the bottom slot in this particular unit. In some units the relationship of these slots is reversed, the long on top and the short on bottom.

Fig. 7-14. A wafer keyway unit (a view of the tumbler side).

Fig. 7-15. A wafer keyway unit (reverse side).

On the reverse side of the keyway unit (Fig. 7-15), notice that there is a metal spring rack (N6) which looks like a comb and upon which are seated the wafer springs (N7).

Keyway Coding

As explained before, the first column to the right of the plunger spring is the master wafer column. Its propor-tions give a clue as to the type of keyway with which we are dealing. In Fig. 7-16A, the longer slot is below the V-grooved dividing strip, indicating a type 2 keyway.

The 14 slots to the right of the master wafer column are assigned code numbers corresponding to the place-ment of the combination wafers (N8). The first slot to the right of the master wafer column above the dividing strip is given the designation 1. The slot directly below this is

Fig. 7-16. The keyway coding system.

71

Fig. 7-17. A keyway unit showing the position of the master wafer in the master column.

Fig. 7-18. A keyway unit showing the positions of the combination wafers for the combination 101450.

given the designation 2. The code numbers alternate between odd and even, continuing to the right of the keyway. All the odd numbers are on top—1, 3, 5, 7, 9, 1′, 3′ (1′ and 3′ are read as 1 prime and 3 prime). All the even numbers are below the dividing strip—2, 4, 6, 8, 0, 2′, 4′.

All factory-cut keys have a combination number stamped on the bow which indicates the notching on the key. In Fig. 7-16B the key has the number 203823. The first digit indicates a type 2 key with its tip cut away. The second digit, in this case 0, indicates a stock key not related to any master keyed system. The last four digits indicate the location of the notches cut in the key. These same four digits (3823) also indicate the position of the combination wafers in the keyway since these wafers must rest within the cutaway portions of the key.

If a key is not stamped with a factory combination number, take an empty wafer keyway and insert the questionable key to determine its combination number. In Fig. 7-16C, the shank of the key may be seen through the slots in the keyway except at those code locations where the key has been notched. Looking first at the master wafer column, the slot below the dividing strip is unobstructed, indicating the tip of the key has been cut away at the bottom. This key is therefore a type 2; the first digit of the combination number would be 2. The second digit of the code number is not related to any cuts on the key and is always 0 for stock (non-master keyed) keys. Next, look at the 14 code slots in the remaining 7 columns to locate the 4 cutaway portions of the key. These cuts occur at code positions 3, 8, 2′, 3′. The complete combination number which would be stamped on this key is 203823.

Keying by Combination Numbers

Using the combination number 101450, the first digit designates a type 1 keyway. Select such a keyway and into this insert a master wafer in the first, or master, column with the protrusion pointing up (Fig. 7-17). The second digit, 0. indicates a stock keyway unit which should be set up in accordance with the following procedure.

Taking four combination wafers, insert them in the positions indicated by the last four digits of the combination (Fig. 7-18). The combination wafer is unique in that it may be inserted with the protrusion pointing either up or down in the seven combination columns. The code number designates in which of the 14 slots the protrusion should be inserted. For example, 1 should be inserted

Fig. 7-19. A keyway unit showing the positions of the series wafers for the combination 101450.

pointing upward in the first column after the master wafer column; 4 should point downward in the second column; 5 should be pointing upward in the third combination column; 0 should be pointing downward in the fifth combination column. After the four combination wafers have been positioned, the remaining three empty columns should be filled with the series wafers (Fig. 7-19). Note that the protrusion of the series wafers can be inserted only in the longer slots of the empty columns. The protrusion of all the series wafers, therefore, should point in the same direction within any one keyway. When springs have been properly attached to all wafers, the keyway unit is then ready to be operated by a key cut to combination 101450.

KEYING ALIKE

Frequently it becomes necessary to alter the combinations of one or more stock keyway units to exactly match those of another. The procedure used to accomplish this is called "keying alike" and should not be confused with "master keying."

One of the simplest methods of keying alike a group of stock wafer keyway units of the same type, either type 1 or type 2, is to put one aside as a control, empty the series and combination wafers from the others, then set up these units to the code combination of the control keyway, using the procedure explained in the preceding section. An alternate method involving fewer operations consists of rearranging only those series and combination wafers in the random keyways which differ in position from those located in the control keyway.

Fig. 7-20. Two keyway units of the same type but with different combinations.

As an illustration of the alternative method, let's key alike two keyway units (Fig. 7-20) of the same type (type 2). We'll use the keyway coded 203823 as the control (A). Of course, we should examine the columns one at a time, moving from the first to the seventh (Fig. 7-21).

First Column   No change necessary (Fig. 7-21A).
Second Column   Invert the combination wafer so the protrusion extends through the code 3 slot (Fig. 7-21B).
Third Column   Replace the combination wafer with the series wafer (Fig. 7-21C).
Fourth Column   No change necessary (Fig. 7-21D).
Fifth Column   No change necessary (Fig. 7-21E).
Sixth Column   Replace the series wafer with a combination wafer. The protrusion must extend through the code 2′ slot (Fig. 7-21F).
Seventh Column   No change necessary (Fig. 7-21G).

CUTTING WAFER KEYS

Keys accurately cut from genuine Schlage key blanks ensure smooth operation if the dimensions of the notching, as shown in Fig. 7-22, are closely observed.

The first operation normally performed on a key blank is to cut away a portion of the tip to correspond to the type of keyway unit with which it is to be used. Key blanks may be purchased with this notch already cut by specifying blanks for type 1 or type 2 keyway units. All other cuts on the key are made to the same depth (0.060″) and have the same width (0.060″) at the bottom of the notch. All the angles in the cuts should be a minimum of 90°.

It is best to use either a factory-cut key blank or a full-cut pattern key available from the factory. With full-cut pattern keys it is necessary to select only those notches corresponding to the specific combination numbers to be cut. After the keys are cut, dress them lightly with a file to remove sharp edges and check the keys in the keyway unit to make sure they operate properly. All the protrusions on the wafers should be flush with the keyway when the proper key is inserted.

ASSEMBLING THE KEYWAY UNIT

As shown in Fig. 7-23A, always hold the keyway unit in your left hand with the keyway cam (N2) to the left and the V-grooved dividing strip facing you. Then as you insert the wafers (B), you will find the series wafers will only go in the elongated slot of the column. They should not be forced because they will not fit into the incorrect position. All wafers should be inserted in the slots with the protrusion first. This protrusion projects between the two side sections of the steel keyway frame.

Fig. 7-21. Keying alike two keyway units.

Fig. 7-22. Standard dimensions of the wafer key.

When all the wafers have been inserted into their proper location, hold the finished cap of the keyway unit with your right hand (C) and move your left hand so that your first two fingers cover the wafers, holding them in position as you rotate the keyway unit (D) to expose the spring rack (N6). If the wafers are not held in position as you rotate the keyway, they may drop out and cause you to rework the setup.

After the keyway unit has been rotated, hold it tightly between your fingers and thumb. Exert force downward against the wafers with the index finger to open the distance between the spring rack in the center of the keyway and the spring seat on the wafers to provide space for the insertion of the wafer springs.

The easiest way to insert a wafer spring is to use a pair of fine needlenose tweezers (E) to grasp the spring at the second coil back from one end. Holding it in this position, you can guide the free end of the spring over the spring rack seat and use the needlenose tweezers to guide the other end over the spring seat on the wafer (F).

Take the key (G) and run it in and out of the keyway several times as it is important that all springs are fully seated before the keyway unit is reassembled into the lock.

MASTER KEYING

Frequently it becomes necessary to provide a group of locks with keyway units, each having a different and noninterchangeable key, but all having in common one master key capable of operating each of the locks in the group.

Combining locks in this manner is called *master keying* and is quite easily accomplished with Schlage

Fig. 7-23. Assembly of the keyway unit.

Fig. 7-24. A list of combination numbers for keys or keyway units.

10	2360
10	3702'
10	1470
10	1360
10	1301'
10	2369
10	1391'
10	2469
10	1791'
10	172'4'

wafers cannot be used in any column for which the key has been notched since it is the unnotched portion of the key which retracts the series wafer. At no time should a keyway unit be set up without any series wafers because this decreases the security of the master key system.

One method of laying out a master keyed system involves five basic steps:

1. Select the required number of keyway units, all of which should be either type 1 or type 2.

2. Of the seven columns available, arbitrarily select either one, two, or three columns which will be reserved in all keyway units for the placement of series wafers. Sixteen master keyed units are available with three columns reserved for series wafers. Eighty master keyed units are available when two columns are reserved for series wafers. Two hundred and forty master keyed units are available when one column is reserved for a series wafer.

3. List the code numbers of all the slots in the remaining columns. These slots are available for the placement of the four combination wafers. These code numbers also represent the cuts in the master key which will operate all the keyway units having combinations determined by the procedure in step 4.

4. Using the digits obtained in step 3, tabulate a list of four digit combination numbers. Although each column contains two code slots, it can accommodate only one wafer. Therefore, be careful to select not more than one digit from any available column.

5. Using the code numbers determined above, key the keyway units by the combination numbers, placing series wafers only in those columns reserved in step 2.

There is an alternate method of master keying which is particularly useful when there is at hand a number of assembled stock keyway units or a supply of factory-numbered stock keys. The following basic steps illustrate this method:

1. Select approximately one-third more keys or keyway units (all of one type) than your anticipated needs

wafer keyway units. To illustrate the principle of master keying, consider two stock keyway units having combination numbers 203823 and 204823. These units, like all other stock keyway units, contain one master wafer, four combination wafers, and three series wafers. Their code numbers indicate that these units are identical except for the position of the combination wafer in the second column. This difference is sufficient to prevent interchangeability of their keys, but a third key could be cut which would operate both these keyway units by providing it with notches for both the code 3 and code 4 tumbler positions. In other words, their master key would be cut with five notches corresponding to code positions 3, 4, 8, 2, and 3.

This principle of master keying can be expanded to include many more combinations. However, as the number increases, it will become necessary to eliminate one or two series wafers from the keyway units. Series

6604430	A
	B
135791'3'	
	A
246802'4'	
3240521	B

Fig. 7-25. A representation of the keyway unit, showing the 14 code slots (A) and the frequency of occurrence (B) of each code number in the list of combinations (Fig. 7-24).

13 791'

246 02'

Fig. 7-26. A master key notched at ten different code locations.

76

Fig. 7-27. A type 1 keyway.

and list their combination code numbers as shown in Fig. 7-24.

2. In this method, only one series wafer is used. So in order to determine which column should be reserved for its placement to best use the maximum number of listed keys or keyway units, represent the keyway unit by drawing a rectangle and indicate the 14 slots by writing in their appropriate code numbers (Fig. 7-25).

3. Above and below the rectangle (opposite each slot code number) write the figure which indicates the frequency of occurrence of the particular code number in the list of combinations.

4. Add the two figures in each column (outside the rectangle) obtained in step 3 and determine the column having the lowest sum. This column will be the best location of the placement of a series wafer in all keyways selected to be master keyed. In Fig. 7-25 the seventh column with a sum of one is the most desirable location for the series wafer.

5. Review the list of combination code numbers and eliminate as impractical all combinations which contain digits corresponding to code slots which are located in the column selected in step 4. In the illustration, combination 101724 would be eliminated since it contains the digit 4', a code slot in the seventh column.

6. The required quantity of keyway units to be master keyed should be selected from the remaining group of combination code numbers. Remove from each of these selected units the two series wafers not located in the column reserved in step 4. Each of these keyway units may still be operated by its own key, but now they all can be operated by a master key.

7. The master key which will operate these units is made by notching the key only at each location where a combination code number occurs. The rest of the key must be left unnotched. The master key for units selected from the list in Fig. 7-24 would be notched at combination code locations 1, 2, 3, 4, 6, 7, 9, 0, 1', 2' (Fig. 7-26).

Master Key Stamping

Some specific designation, such as the letter M, should be stamped on the bow of master key. Factory-cut master keys are stamped with a registry number prefixed with the letter R. This number is indexed in the factory records, which contain detailed information such as the job name, the job location, and the combinations controlled by the master key. Factory-cut change keys, which operate the individual keyway units in a master keyed system, are stamped with a combination number. This number has the same meaning as that stamped on stock keys. The second digit of the master key code

135791 '3'

246802 '4'

Fig. 7-28. In this keyway the seventh column is reserved for a series wafer.

Fig. 7-29. One slot in this column is reserved for each master keyed group.

number is never 0 as it is on stock keys. It represents the number of the column in the keyway which contains the series wafer. Stock keys converted into master keys should be overstamped with the appropriate second digit.

Occasionally it becomes necessary to provide two or more groups of master keyed keyway units (each group having its own master key) with a key called a grand master key which would operate the keyway units in all the groups. One method of laying out a grand master key system involves seven basic steps:

1. Select either type 1 or type 2 keyway with which to build the system (Fig. 7-27).

2. Of the seven columns available, arbitrarily select one column which will be reserved in all keyway units for a series wafer (Fig. 7-28).

3. If only two master keyed groups are desired under the grand master key, arbitrarily select another column which will provide two slots, each of which will be assigned exclusively to a master keyed group and will be the means for separating the two groups, thus preventing interchangeability of their keys (Fig. 7-29).

4. List the code numbers of the slots in the remaining five columns.

5. Tabulate a list of four digit combination num-

1359	3579
1369	3589
1379	3591'
1389	3592'
1391'	4579
1392'	4589
2359	4591'
2369	4592'
2379	4791'
2389	4792'
2391'	5791'
2392'	5792'

AND SO FORTH CONTINUING TO A MAXIMUM OF 80 COM-BINATIONS   135791'

2468 2'

Fig. 7-30. A list of combination numbers for a master keyed group. Each number contains the unique slot code number 9.

1350	3570
1360	3580
1370	3501'
1380	3502'
1301'	4570
1302'	4580
2350	4501'
2360	4502'
2370	4701'
2380	4702'
2301'	5701'
2302'	5702'

AND SO FORTH, CONTINUING TO A MAXIMUM OF 80 COMBINATIONS

1357 1'

246802'

Fig. 7-31. A list of combination numbers for a master keyed group. Each number contains the unique slot code number 0.

Fig. 7-32. A grand master key that fits all the keyway units in two master keyed groups (Figs. 7-30 and 7-31).

Fig. 7-34. Assembly of Schlage wafer lock, Step 1.

bers, each of which must contain one of the unique slot code numbers selected in step 3. The remaining three digits may be drawn from the available code numbers obtained in step 4, but remember to select not more than one digit from any one available column. This list of combination numbers represents one master keyed group (Fig. 7-30).

6. Repeat step 5. However, substitute the other unique slot code number for the one previously used. This list of combination numbers represents the second master keyed group. Though these two groups may appear similar, their master keys are not interchangeable since one will not have the notch which is necessary to operate that combination wafer which is used exclusively in the other group (Fig. 7-31).

7. The grand master key has all the notches found on both master keys and will, therefore, operate all of the keyway units in both groups (Fig. 7-32). This illustration represents only one of a large variety of possible grand master keying arrangements. More complex systems or those involving a greater number of master keys should be referred to the factory.

Fig. 7-35. Assembly of Schlage wafer lock, Step 2.

Fig. 7-36. Assembly of Schlage wafer lock, Step 3

Fig. 7-33. Two keyway units (type 3 and type 4).

Extended Master Keyed and Grand Master Keyed Systems

Master keyed or grand master keyed systems may be doubled simply by using both types of keyways (type 1 and type 2) and providing a key capable of operating both. In such an extended system, all keyways must be provided with a series wafer in the master column. This does not, in any way, impair the operation of any existing keys. However, it does make possible the use of an additional

Fig. 7-37. Assembly of Schlage wafer lock, Step 4.

Fig. 7-39. Assembly of Schlage wafer lock, Step 6.

Fig. 7-38. Assembly of Schlage wafer lock, Step 5.

Fig. 7-40. Assembly of Schlage wafer lock, Step 7.

key (type 0) capable of acting as a master, grand master, or great grand master key. When a type 1 keyway unit is set up with a series wafer replacing the master wafer, both the keyway unit and its key are redesignated as type 3. Similarly, a type 2 keyway unit and its key are redesignated as type 4 (Fig. 7-33).

REASSEMBLING THE LOCK MECHANISM

1. Refer to Fig. 7-34. Grasp the wafer keyway unit (N) in your right hand. In your left hand, hold the outside knob (M) with the slot (M1) facing you. Insert the wafer keyway unit (N) into the outside knob spindle aligning the keyway cam as shown. Depress the wafer protrusions between your thumb and forefinger, pushing in the keyway unit until it is bottomed.

2. Insert the knob through the outside rosette and frame (Fig. 7-35), pushing and rotating simultaneously until it is fully assembled. The keyway cam and spindle ears should be positioned at the open end of the lock frame.

3. Refer to Fig. 7-36. Place the springs (K) into the slide (J) and assemble both into the frame, positioning the free ends of the springs on the ears (H2) of the lock frame. Be sure the bridge of the slide is facing you, as shown.

4. Refer to Fig. 7-37. Replace the thrust plate (I) with its curved side over the springs (K) by inserting the bottom side between the bottom tabs of the lock frame and pressing the snap into position between the top tabs (H1).

5. With your thumb depress the slide all the way back against the spring tension (Fig. 7-38). Insert the empty spindle (E) as shown. Release the slide so the spindle will lock into position.

6. Refer to Figs. 7-39 and 7-40. Slip the plunger (L) inside the empty spindle, making sure that the tab (L1) on the turn button aligns with the locking slot (E1). Again depress the slide and work the plunger in all the way. Release the slide so the plunger will lock into position.

7. Replace the housing over the lock assembly with its opening in line with the jaws of the slide, and replace the cotter pin. The lock is now ready for installation in the door.

Chapter 8

Pin-Tumbler Cylinder Locks and Locksets

Pin-tumbler cylinders are basically very simple mechanisms but can be built to give a high level of security.

CONSTRUCTION

The cylinder is a self-contained mechanism, intended to be adapted to a variety of locksets. Figure 8-1 shows a pin-tumbler cylinder in cross section.

Plug

The plug contains a keyway, usually of the *paracentric*, or off-center, type. It is drilled for four or, in some cases, as many as seven tumbler chambers. These chambers are made with the greatest precision. Chambers are spaced evenly along the upper surface of the plug and are aligned as closely as modern production techniques allow. The cylinder shell, sometimes called the *hull*, carries a matching set of chambers.

The plug may be machined with a shoulder at its forward surface; this shoulder mates with a recess in the cylinder and provides:

☐ A reference point for regulating the alignment of the pin chambers in the shell and the plug.

☐ A safeguard to prevent the plug from being driven through the cylinder, either deliberately or through resistance developed as the key enters the plug.

☐ A safeguard to discourage a thief from shimming the pins. Without this shoulder it would be possible to force the pins out of engagement with a strip of spring steel.

The plug is retained at the rear by a cam and screws, a retainer ring, or a driver pin that locks the plug into the cylinder. The retainer pin is mounted securely in the plug and fits into a groove milled on the inside diameter of the cylinder shell. This groove determines the degree of plug rotation. The groove can be milled to stop plug travel when the bolt is thrown; the key can be withdrawn only if the bolt is in the locked position. Without this travel-limiting feature, the cylinder can be rotated to release the key in the unlocked position.

Pins

Pins are supplied in a variety of sizes and shapes—truncated, truncated with modified cone, hemispherical, capped, or any of a number of other designs. Figure 8-2 shows some of the more popular configurations. The drawings have been exaggerated for clarity.

Fig. 8-1. Yale pin-tumbler cylinder in cross section.

The shape of the pins can foil would-be lock pickers. A standard straight-sided cylindrical pin can easily be raised to the shear line while the plug is turned slightly. A pin with a broken profile may tend to "hang up" before it passes the shear line. In this case, only the proper key would align the pin at the shear line.

Pins also come in varying lengths and diameters, depending upon the size of the pin chamber diameter and whether the lock can be master keyed.

Although variations in pin length and diameter seem insignificant, these small differences can be critical. For example, a pin that is a few hundredths of an inch smaller than the chamber diameter can bind in the plug or cylinder shell. In extreme cases, the pin may jam so solidly that the cylinder will have to be broken in order to open the lock.

Pins are color-coded according to length. The code differs among manufacturers, but each of the seven pin lengths has its own distinctive color.

Cylinder Bars

The typical rim pin-tumbler cylinder has a flat cylinder bar, or tailpiece, attached at the rear of the cylinder plug (Fig. 8-3A and B). The attachment, or tailpiece, is loose to allow some flexibility in the location of the auxiliary lock on the other side of the door. Alignment, while not absolutely critical, should be as accurate as possible; under no circumstances should the tailpiece be more than ¼" off the axis of the plug.

On the other hand, mortise-lock cylinders for mass-produced doors do not have this tolerance. The best is driven by a cam on the back of the lock (Fig. 8-3C). If the lock is used on office equipment, the cam is a milled relief on the back of the plug, or else it is a yoke-like affair, secured to and turning with the plug. It is important that these locks are aligned with the bolt mechanism.

OPERATION

When the correct key is inserted, the key cuts align the tumblers as shown in Fig. 8-1. If the key is cut improperly, the break between the upper and lower pins will not align with the shear line, and the plug will not turn. Better pin-tumbler locks will require a key cut to a tolerance of ±0.025".

DISASSEMBLY

Disassembly of a plug cylinder is simple, requiring only a few basic tools. You will need a screwdriver, a following plug of the correct diameter, and a pin tray. Proceed as follows:

1. Remove the cam or tailpiece.
2. Turn the key in the plug about 30°.
3. Holding the plug so the pins are vertical, slip the appropriate plug follower into the cylinder from behind.
4. Push the plug follower all the way over the plug. Remove the cylinder.

Fig. 8-2. Various pin tumblers.

Fig. 8-3. Medeco cylinders. The tailpiece may be serrated (A), solid (B), or replaced by a cam (C).

5. Taking the plug in hand, slide the follower past one pin chamber; turn the plug over and allow the pin to drop into your hand or on the work table. Place the pin into the first compartment of the pin tray.

6. Uncover the next plug chamber and place the pin in the adjacent compartment of the tray.

7. This step involves removing the upper pins and springs from the shell. Invert the shell. Slide the plug follower back. The pin and spring will drop out as the follower uncovers the No. 1 chamber. Place these parts in the No. 1 pin tray slot above the lower pin. Proceed to the other chamber.

ASSEMBLY

Assembly, or reassembly, of a cylinder from scratch is the reverse of this procedure. You will need pin tweezers to install the upper pins, springs, and lower pins into their chambers. You will need a plug follower to keep the parts together. A plug holder is optional; it holds the plug in an upright position and leaves your hands free while you insert the lower pins.

This procedure assumes that you have a key but do not know the pin combinations. (In normal reassembly, you already know which pins go into which chambers.)

1. Place the plug into the plug holder and insert the key.

2. Taking one lower pin at a time, insert it into the cylinder chamber. If a pin stands above or below the shear line, another must be tried. If the pin looks like it is just about at the shear line or perhaps right on it, turn the plug with the key. If the plug turns, you know that the pin is the correct one for the chamber.

3. If the pin is the wrong one, remove it and select another. Once the proper pin is located, leave it in the chamber and move on to the next one. Repeat this proce-

dure until you have the proper pins for all the chambers. *Note: Do not remove the key or turn the plug over.*

4. Move to the cylinder shell and insert the plug follower.

5. The plug follower has a notch cut into its end; keep this notch aligned with the chambers. Start at the third chamber from the front.

6. Using your tweezers, select a spring and place it into the loading notch of the follower. Rotate the cylinder so that the spring slides into the chamber. Select an upper pin (all upper pins are the same size) and, using the tweezers, place it into the loading notch.

7. Push the pin against the spring and into the pin chamber. At the same time, slide the plug follower forward. This movement holds the pin in No. 3 chamber and uncovers No. 2 chamber.

8. Load No. 2 chamber as before and proceed to No. 1 chamber.

9. Once the first three chambers are filled, retract the follower, leaving the last two chambers exposed (if you push the follower too far, the pins in the front chambers will pop out).

10. Load No. 4 chamber and then No. 5. Once the chambers are loaded, you are ready to insert the plug into the cylinder.

11. Place the plug against the end of the plug follower with the assembly positioned in front of the cylinder. Rotate the plug to bring the pins about 30° off the vertical; press against the plug follower tool as you push the plug home. The plug forces the tool out the rear of the cylinder. Since the plug and plug follower are butted, there is no opportunity for the upper pins to fall out.

12. Once the plug is home, hold it in place and turn the key. If all is well, the plug will turn smoothly. If it binds, disassemble the lock and check the pin lengths.

13. Attach the cam or tailpiece and retest.

CHOOSING A PIN-TUMBLER CYLINDER LOCKSET

Pin-tumbler cylinder locksets are the most popular type of locking device. Since these locksets are available in a wide variety of types and styles, some discussion of their general characteristics is in order.

Security—The pin-tumbler cylinder lockset provides better than average security by virtue of its design. However, security also depends upon the quality of the lock and upon its application. Metal doors give more protection than wood doors.

Quality—Quality depends upon the intended service of the lockset. Light, medium, and heavy-duty sets are available.

Type—Locksets are identified by their function—lavatory, classroom, residential, and so on.

Visual appeal—The lockset should match the decor of its surroundings. This is particularly important in new constructions.

Hand—The location and direction of swing of the hinges determines the *hand* of the door (Fig. 8-4). Taking the entrance side of the door as the reference point, there are four possible hands: left-hand, left-hand reverse (the door opens outward), right-hand, and right-hand reverse.

LEFT HAND: HINGES ON LEFT, OPENS INWARD.

RIGHT HAND: HINGES ON RIGHT,

LEFT HAND REVERSE: HINGES ON LEFT, OPENS OUTWARD.

RIGHT HAND REVERSE: HINGES ON RIGHT, OPENS OUTWARD.

Fig. 8-4. The *hand* of a door is a term that describes the location of the hinges and the direction of swing. (courtesy Eaton Corp.)

A – Deadbolt	**J** – Outside Hub	**T** – Latch Lever Spring
B – Deadbolt Spring	**K** – Inside Hub	**U** – Auxiliary Latch
C – Turn Knob Hub	**L** – Short Stop Button	**V** – Deadlocking Lever
D – Latch Lever	**M** – Long Stop Button	**W** – Deadlocking Lever Spring
E – Latchbolt regular	**N** – Button Cam	**X** – Stop Cylinder Lever
F – Long Hub Lever	**O** – Button Cam Spring	**Y** – Cylinder Latch Lever
G – Short Hub Lever	**P** – Latch Lever Spring	**Z** – Stop Lever
H – Hub Spring	**S** – Auxiliary Latch Lever	**AA** – Anti-Friction Latch Bolt

PARTS NOT SHOWN

BB – 1300N Hub

DD – 1300E Latch Lever

Fig. 8-5. Pin tumbler cylinder mortise lock parts identification. (courtesy Dominion Lock Co.)

Fig. 8-6. Corbin Model 68 lockset in cutaway.

It is important to match the lockset to the door hand. Failure to do so can cause bolt/striker misalignment and may require that the cylinder be rotated 180° so that the weight of the pins is on the springs. The pins weigh only a few grams, but this is enough to collapse the springs and disable the lock. Unless stipulated in the order, manufacturers supply right-hand mortise locksets. Some of these locksets can be modified in the field; others are fixed.

The mortise lock is a standard one, used in homes, apartments, general business and large industry, in addition to government and private institutions. Extremely popular, it is standard fare for the locksmith. Figure 8-5 shows the two basic mortise lock models available and in great use. Although the individual parts may, at times, vary slightly from manufacturer to manufacturer, they are essentially the same. You should study this picture, then take apart one or several of your mortise locks in stock and study the parts and their relationship to other parts. Because of the abundant use of these locks, you should know these, like other standard locks, from memory.

SPECIFIC LOCKSETS

The information presented in the following pages has been supplied by the various manufacturers. While it does not begin to cover *all* locksets, it does give an overview of the current state of the art. If you need further information, contact one of the factory representatives.

Corbin

Figure 8-6 is a sectional view of the popular Corbin Model 68 lockset. This lockset gives adequate security for most homes and apartments. Model 323D is intended for medium and heavy-duty applications and features a positive thumbturn (Fig. 8-7). Model 354 is a superior lockset intended for severe duty (Fig. 8-8). Parts are made of high-quality steel; they are stronger than they need to be.

Corbin's unitized lockset has some unique features (Fig. 8-9). The knob is mounted on a full-width bearing for better wear characteristics; the swinging latchbolt ensures positive locking action; the interior parts are bronze plated to prevent corrosion and to reduce friction; retractor springs are made of phosphor bronze and are stressed in compression to reduce fatigue. Other features include an optional extended-lip escutcheon that guards the bolt and the deadlocking latchbolt (Fig. 8-10).

Mounting the lock is simple. All that is required is a rectangular cutout from the door edge and a pair of holes for the through bolts.

The Corbin Fortune 1400 is one of the most interesting locksets on the market (Fig. 8-11). This lockset is priced competitively but is designed to give long service. The bolt mechanism has only four moving parts and the knob is super rigid. The cylinder can be changed in a matter of minutes, a feature that should reduce maintenance costs.

122 FOR No. 322D ONLY

1 2 3 76 66 22 116 5 123 10 124 53 110 12 14   114 FOR No. 323D ONLY

B

38 46 19 120 32 FOR No. 301D ONLY

6
24

SECTION "B," "B"

24

126

37

ASSEMBLY OF
LATCH TO SHOE

12 23 5   12 23 5

INSIDE   OUTSIDE   INSIDE   OUTSIDE

1⅜ in. REVERSE STRAIGHT FACE   1¾ in. REVERSE BEVEL FACE

5 23 12   5 23 12

OUTSIDE   INSIDE   OUTSIDE   INSIDE

1⅜ in REGULAR STRAIGHT FACE   1¾ in. REGULAR BEVEL FACE

⅝″ THROW USED IN 1⅜″ FRAME ¾″ THROW USED IN 1¾″ FRAME

Fig. 8-7. Corbin Model 323D lockset with dead bolt.

ASSEMBLY
OF LATCH
TO SHOE

ASSEMBLY
OF TRIP LATCH
TO LATCH TAIL

SECTION "B", "B"

Fig. 8-8. Corbin Model 354 is intended for severe duty.

Corbin's Ensign series is reversible; it can be used on left or right-hand doors. However, these locksets do not have an auxiliary bolt (Fig. 8-12).

Schlage "A" Series

Schlage engineers have worked for many years to perfect the cylinder lockset. These efforts, together with rigid quality-control standards, have made Schlage a leading name in locks.

"A" series standard duty locks are recommended for normal use applications; "G" series heavy-duty locksets combine a deadbolt with a deadlatch for additional security and durability.

Figure 8-13 is a cutaway view of the "A" series lockset. As shown, the cylinder has five pin tumblers; an additional pin is optional. The pins are encased in an extruded brass cylinder plug and shell. These locks are recommended for residential and commercial buildings where the normal frequency of operation is anticipated, and are used in motels, office buildings, warehouses, and retail stores, as well as in apartment complexes, residential homes, medical clinics, schools, and churches. Unlike

NYLON ADJUSTING
SCREW

ASSEMBLY
THROUGH BOLT

DUST-PROOF
FRAME COVER

SOLENOID
FRAME

BOX STRIKE

LATH BOLT

AUXILIARY LATCH

Fig. 8-9. Corbin's unitized lockset is becoming increasingly popular.

Fig. 8-10. Extended lip escutcheons limit access to the latchbolt. (courtesy Emhart Corp.)

many other locksets on the market, the Schlage "A" series can be adapted to meet almost any installation requirement. An extension link can be mounted on the latch, and the knob can be positioned as far back as 10 inches from the door edge. The backset— the distance from the front of the lock to the center of the cylinder hole or knob hub—is 5 inches (Fig. 8-14).

A51DP—The A51DP is an entrance lock (Fig. 8-15). Rotating either knob will retract the latchbolt. The turnbutton is on the inside knob; moving it to the horizontal position locks the outside knob. It must be released manually and restored to the vertical position; otherwise, only the key will unlock the lock. The inside knob is always free for immediate exit. The latchbolt automatically deadlocks the door when it is closed.

A73PD—Rotating either knob will retract the latchbolt; pushing the button on the inside knob will lock the bolt; from the outside, the bolt is key actuated (Fig. 8-16). Turning the inside knob or closing the door releases the button to prevent accidental lockout. Lubrication points are indicated in this and the previous figure by arrows. While these locks share many of the same parts, the A73PD has a few more for better security.

Schlage "G" Series

"G" series double-locking security locksets are designed for high security and long wear. The "G" series combines a 1″ throw deadbolt—with a roller tip—and a ½″ throw deadlatch. The lock case is backed up with armor plate, and the internal mechanism is made of cold-rolled steel chemically treated to resist corrosion. When locked, the outside knob spins freely to discourage forced entry. Coil compression springs are used exclusively. In addition, these locks have *six* pin tumblers set into an extruded brass cylinder. The pins are made of nickel-silver to ensure long wear.

The lockset has been designed for easy installation; all that is required is to drill five holes in the door. Friction-grip screws secure the lockset to the door. While the hand can be reversed in the field, it is simpler to order a lockset with the correct hand. The standard backset is 2¾″. Backsets up to 5″ can be ordered.

G51PD—The G51PD is illustrated in Fig. 8-17. It has these characteristics.

☐ Rotating either knob retracts the latchbolt.
☐ Turning the button in the inside knob to a horizontal position disengages the outside knob and makes it inoperative.
☐ The button does not release unless manually restored to the vertical position.
☐ Rotating the thumbturn from the inside, or the key from the outside, extends the deadbolt to the locked position.

Fig. 8-11. Corbin Fortune 1400.

Fig. 8-12. Corbin Ensign.

Fig. 8-13. Schlage series lockset.

☐ Both the deadbolt and the latchbolt can always be retracted to the unlocked position by rotating the knob or the thumbturn from the inside or the key from the outside.

☐ The latchbolt automatically deadlocks when the door is closed.

The dual-cylinder version of this lockset is used almost exclusively in hotels and motels. A plastic cylinder cover blocks access to the upper cylinder. The cover can easily be removed with a special extractor tool. For extra security, the upper and knob cylinders can be fitted with different keyways. The change key (the one held by the guest) retracts the bolt; the deadbolt is extended or retracted by the emergency key. The inside knob or the

Fig. 8-14. Critical lockset dimensions. (courtesy Eaton Corp.)

**USE LUBRIPLATE M-204-168 (130-AA)
WHERE INDICATED BY ARROW**

Fig. 8-15. Schlage A51DP lockset in exploded view. Arrows indicate lubrication points.

Fig. 8-16. Schlage A73PD lockset in exploded view. Arrows indicate lubrication points.

SLIDE UNIT

G505-748

Fig. 8-17. Schlage G51PD lockset.

92

Fig. 8-18. Lubrication points for the Schlage G51PD.

thumbturn retracts both bolts simultaneously for immediate exit. The latchbolt automatically deadlocks when the door is closed.

Figure 8-18 indicates the lubrication points for the "G" series lockset. Use graphite in the cylinder and Lubriplate M204-168 on the other parts.

Medeco

Many locksmiths believe that the Medeco cylinder is the ultimate in pin-tumbler security.

Operation. Figure 8-19 is an exploded view of a typical Medeco six-tumbler cylinder. The pin tumblers must be raised to the shear line between the plug and cylinder shell by the key cuts. The depth of the individual cuts corresponds to the length of the tumblers. This principle is used on all pin-tumbler locks built on the Yale pattern, but Medeco has added another line of defense— one that makes key duplication almost impossible.

The tumblers, responding to angled bites in the key, rotate as they raise. The tumblers are slotted and, in the unlocked position, align with extrusions on the fence or sidebar. Tumbler slots and the fence are shown as parts Nos. 11 and 4 in the exploded view. Figure 8-20 shows the slot for the sidebar on the plug. A matching slot is milled on the cylinder shell. In the locked position, the sidebar is athwart these slots, effectively blocking plug rotation (Figs. 8-21 and 8-22). In the unlocked position, the tumblers have turned so that their slots are at right angles to the blade of the key (Fig. 8-23). When this condition is met, the sidebar is free to move into the plug. If the tumblers are at the correct height, the lock will open (Figs. 8-24 and 8-25). Figure 8-26 is a closeup of the engagement mechanism.

The key is a very special item (Fig. 8-27). Angles milled on the sides of the bites determine the direction and amount of pin rotation and add another level of complexity. There are some 23 million usable combinations.

ASSEMBLY NUMBER 10-200—MORTISE CYLINDER			
ITEM	DESCRIPTION	REQ'D	PART NO.
1	Key 6 Pin	1	10-011
2	Plug Mortise 6 Pin	1	10-012
3	Shell Mortise 6 Pin	1	10-017
4	Fence 6 Pin	1	10-022
5	Washer	1	10-025
6	Cam	1	10-027
7	Spring Cover 6 Pin	1	10-030
8	Screw # 2-56 Slotted F. H.	*2	10-032
9	Spring "Tumbler"	6	10-034
10	Driver (Hardened on application)	6	10-035—10-040
11	Tumbler	6	Listed below
12	Spring Fence	2	10-047
13	Master Pin Disc	As required	10-048—10-052

ASSEMBLY NUMBER 10-400 RIM CYLINDER			
ITEM	DESCRIPTION	REQ'D	PART NO.
1	Key 6 Pin	1	10-011
2	Plug Rim 6 Pin	1	10-020
3	Shell Rim 6 Pin	1	10-024
4	Fence 6 Pin	1	10-022
5	Retainer	1	10-026
6	Connecting Bar	1	10-028
7	Spring Cover 6 Pin	1	10-030
9	Spring "Tumbler"	6	10-034
10	Driver (Hardened on application)	6	10-035—10-040
11	Tumbler	6	Listed below
12	Spring Fence	2	10-047
13	Master Pin Disc	As required	10-048—10-052
14	Mounting Screws	2	10-055
*15	Steel Insert (Hardened)	2	10-057
*16	Steel Ball (Hardened)	1	10-058
*17	Escutcheon Ring	1	10-059
*18	Security Pin	2	10-062
*19	Security Pin	1	10-063
20	Plate (Back)	1	10-064
*21	Mushroom Drivers	As required	10-065—10-071

*These parts included on 10-200 MORTISE CYLINDER
Patent No. 34.99302

TUMBLERS		
1	Right	10-041
2	Right	10-042
3	Right	40-133
4	Right	40-134
5	Right	40-135
6	Right	40-136
1	Center	40-141
2	Center	40-142
3	Center	40-233
4	Center	40-234
5	Center	40-235
6	Center	40-236
1	Left	40-241
2	Left	40-242
3	Left	40-333
4	Left	40-334
5	Left	40-335
6	Left	40-336

Fig. 8-19. The Medeco cylinder.

Fig. 8-20. Cutaway showing the pin chambers and sidebar slot.

An ordinary key machine cannot duplicate a Medeco key. Machines are available through the factory, but they are expensive and distribution is rigidly controlled. Only well established locksmiths need apply.

Additional security is provided by tool-steel se-curity pins in front of the tumblers and by armored inserts flanking the keyway and covering the pin engagement slots (Fig. 8-28). The sidebar is protected by a hardened steel ball.

Applications. Medeco cylinders are available as

LOCKING SLOT
IN CYLINDER

SIDE BAR
IN LOCKED POSITION

Fig. 8-21. In the locked position the sidebar extends into the cylinder shell slot.

Fig. 8-22. This drawing shows both locking arrangements: the sidebar and the pin tumblers lie across the shear line.

part of complete locksets or can be ordered to fit competing brands. Cylinders can be adapted to locks made by Adams Rite, Arrow, Corbin, Falcon, Keil, Russwin, Segal, Sargent, Schlage, and Yale.

Service. The people at Medeco have an overriding concern for security. Locks and keys are sent by registered mail. Only well-known and trusted locksmiths are involved in the repair and maintenance of these locks,

TUMBLERS
PROPERLY ALINGED
FOR ENTRY
OF SIDE BAR

Fig. 8-23. The key rotates the tumblers, bringing the slots into alignment with the sidebar.

AS KEY IS
REMOVED
TUMBLERS
ROTATE TO
LOCKED
POSITION

SIDE BAR
ENGAGES SLOTS
ON ALL TUMBLERS
IN UNLOCKED
POSITION

Fig. 8-24. Once the pins are in alignment, the sidebar retracts into the plug.

since parts availability and service information is rigidly controlled by the factory.

Dominion

Dominion Lock of Canada has for many years pro- vided a wide assortment of locking devices and lock-smithing accessories. A standard unit used throughout all of North America is a mortise lock. The Dominion model is referred to as the Dynabolt. It is a heavy-duty mortise delock, in the 370 series (Fig. 8-29).

Fig. 8-25. In the unlocked position the pins and sidebar are below the shear line.

Fig. 8-26. A closeup of the sidebar and pin mechanism.

The unit is for use in doors 1¾" thick and over. The case is pressure cast and has no projections which assures a snug and solid fit. It also means interchangeability with another unit is assured with no problems arising in fitting the unit into the mortised door.

An armored front, 1 1/16" × 4 13/16" can be adjusted from a flat to a standard bevel (⅛" in 2"), and also to any hand required. The unit backset is 2¾".

Unlike many other mortise deadbolts on the market, this particular lock has a pressure cast deadbolt with a 1" throw, but also includes an all-important additional security protection feature: two free-turning hardened steel

Fig. 8-27. Carefully milled angles on the key bitting rotate the pins.

Fig. 8-28. The shaded parts are hardened to resist drilling.

inserts, which provide a greater level of protection against the deadbolt being cut by a hacksaw.

Figure 8-30 provides the lock unit specifications. Positioning of the various trim for this deadlock and identification of the various lock trim parts is shown in Fig. 8-31.

Lock cylinders to be used with the Dynabolt are Sargent, Corbin, Russwin, Falcon and Schlage.

Kaba Security Locks

A new member of the lock industry is Kaba Locks, Ltd., of Great Britain. Kaba locks are not new to most of the world, but they are definitely new to the United States. A major producer with a U.S. outlet, Lori Locks Corporation, Kaba has already made great strides in lock technology. Their locks are becoming more and more widely accepted and used on a worldwide basis. Why? Because Kaba has an ingenious lock, having over 100 million permutations, which makes the lock an excellent one when master key systems are concerned. The most noticeable feature of the Kaba lock is the design of the key (Fig. 8-32).

Principle of operation. The lock has two main parts, a *stator body*, and a *slotted rotor* into which the key is inserted. Aligned holes are machined into the stator and

rotor to take an assembly consisting of tumblers, caps, and springs which are held in place by spring cover plates dovetailed into place (Fig. 8-33).

Figure 8-34 shows a simple group of five such as-

Fig. 8-29. Mortise deadlock with hardened steel inserts in the bolt. (courtesy Dominion Lock Co.)

Fig. 8-30. Dynabolt lock specifications. (courtesy Dominion Lock Co.)

semblies before the key is inserted. Note that all the tumblers project into the slot, under the action of the springs. If the key is inserted, as in Fig. 8-35, in the direction of the arrowhead from right to left, it will obviously lift each tumbler by an amount depending on the depth of the dimple in the key at any point. Clearly, if the depth of the dimple is such that the top of each tumbler just coincides with the interface A between the rotor and stator, the key can be turned and the lock opened. Of course, all five tumblers have to be lifted exactly the right amount for this to be possible.

There are two rows of dimples (A) and (B) on each face of the key. So that the dimples (A) on one face, and the dimples (B) immediately opposite on the other face can pick up at the same time with their tumblers, the key is offset from the centerline (ZZ) of the lock, as indicated in Fig. 8-35. One great feature of the key is that it can be inserted either way round, as one side is just the reverse of the other.

In an actual lock, the dimple depths are not all the same. There are four different values to choose from, referred to as 1, 2, 3, or 4, the deepest being 1 and the shallowest 4. For each depth, there is a particular tumbler length, cap and spring, and the lock is assembled according to a carefully worked out permutation board, on which particular assembly sequences, such as 21212 or 42441, etc., are listed from the terrifically wide range of possible depth combinations.

The lock shown in Fig. 8-36 is referred to as the Kaba 20, as it can have a maximum of 20 tumblers (as opposed to five, six or seven normal tumblers for the average pin tumbler American lock), arranged as shown in Fig. 8-37, with five holes drilled on (A), five holes diametrically opposite on (b), five holes on (C) at 45 degrees to (A) and five holes on (D) at 45 degrees to (B). Some of the locks in the Kaba 20 range, however, use only twelve of the possible 20 tumblers, disposed as in Fig. 8-38, according to the permutation being used.

EE — Inside Knob
GG — Inside Rose & Ferrule
HH — Inside self-aligning Assembly with Oval Head 1 x 8/32 Attaching Screws
JJ — Outside self-aligning Assembly
KK — Outside Rose & Ferrule
MM — Outside Knob
NN — Front Plate Screws — Flat Head Undercut 1/4 x 32
TT — Thumb Turn & 3/8 x 6 Oval Head Screws

Fig. 8-31. Dynabolt mortise lock trim and parts identification. (courtesy Dominion Lock Co.)

Fig. 8-32. The uniquely designed Kaba key provides exceptionally high security integrity. (courtesy Kaba Locks, Ltd.)

The body and rotor are both made from solid brass and completely machined; no castings at all are used. The keys, of a nickel alloy, are fine-blanked and then milled to size to a tolerance of one-thousandth of an inch on thickness. The width and thickness are both most important for the successful operation of the lock. The appropriate dimple depths are obtained in a drilling machine especially developed for this purpose. All parts, including the nickel alloy tumblers, are extremely precisely made.

Kaba Locking System Levels. Kaba locks provide for three levels of security protection. First, there is

Fig. 8-33. Exploded view of Kaba 20 high security lock. (courtesy Kaba Locks, Ltd.)

SECTION ZZ LOOKING INTO THE
SLOT FROM THE LEFT

Fig. 8-34. Cutaway showing front and side views of a Kaba lock. (courtesy Kaba Locks, Ltd.)

the Kaba 20 system, which provides high security, control, and convenience. It provides for over 100 million different key combination possibilities. Individual key registration with control over the supply of duplicate keys, also provides for duplicate keys made against an authorized signature. A wide range of different lock types are available within the key system which eliminates large bunches of assorted keys to be carried by an individual. This system also enables a very close control over access, improving both security and safety. Finally, the extensive master key systems that are available can be individually designed to suit any application (airports, hotels, hospitals, banks, factories, nuclear facilities, etc).

The second level of security protection is provided by the Mini-Kaba system, which also provides wide versatility, control, and convenience in its system. The Mini-Kaba allows for over one million different key combinations. The cylinders are available to suit many existing cylinder mortise locks, in an extensive dimensional capability, available in cylinder sizes down to 16mm diameter. This system also has the master key capability, and further allows for key machines available to local locksmiths for key service.

The final system is the MicroKaba system which gives miniaturization, control, and convenience to the keyholder. More than 10,000 different key combinations are available. The minute 12mm cylinder can be installed in the smallest equipment or components. These are available in a wide range of key-operated switches (Figs. 8-36, 8-39 and 8-40 illustrate these three lock types).

Security Key Systems. The three systems available (central locking, master key, and grand master) are available to the locksmith from Kaba Locks. The central locking system (Fig. 8-4), is really designed for multiple

Fig. 8-35. Front end cutaway view with key inserted. (courtesy Kaba Locks, Ltd.)

Fig. 8-36. Kaba 20 cylinder cutaway; notice the four planes for the cylinder pins. (courtesy Kaba Locks, Ltd.)

dwellings in which several doors are used by all personnel. As you can see from the illustration, a small business could also use such a system with great efficiency and still retain a high level of security integrity. The doors with a central lock and others (such as side entrances, storage rooms, and possibly a garage) can be opened by all. Other doors can also be put on the system if so desired.

The master key system (Fig. 8-42) can also be used

for a variety of applications from the home to medium-size industrial and administrative buildings. Using a single key (the master key) all doors are opened. The individual key opens only the lock for which it was specifically designed. Rooms or even cupboards which are the same classification within the organization can be fitted with locks of the same key combination, if desired.

The grand master key system (Fig. 8-43) is particu-

Fig. 8-37. Diagrammic front view showing the pin alignment. (courtesy Kaba Locks, Ltd.)

Fig. 8-38. Using only 12 pins, two pin chamber rows carry only one pin each, while the other two chamber rows have a full complement of five pins. (courtesy Kaba Locks, Ltd.)

Fig. 8-39. Cutaway view of the Mini-Kaba lock. (courtesy Kaba Locks, Ltd.)

Fig. 8-40. Micro-Kaba switch lock. On the right is the core removed. The switch size (compare with coin) provides for exceptionally wide varieties of uses with electrical and electronic circuitry. (courtesy Kaba Locks, Ltd.)

Fig. 8-41. Central locking system. (courtesy Kaba Locks, Ltd.)

larly suited for larger companies in which staff responsibility is limited to a single or series of particular areas. The most complex and diverse locking arrangements can be undertaken using a Kaba grand master system. An individual key can fit all the locks, or only one door, or all doors of a particular department, or only the corridor doors, or even only all the workshops or offices.

The Kaba 20 and Mini-Kaba are only part of a comprehensive and highly specialized range of products that are available. On the following pages you will find specific locks that can provide for an extremely wide and diverse range that have benefit for your potential customers.

Mortise Lock Cylinders. The 22mm profile cylinder for the mortise lock comes in several variations. The

Fig. 8-42. The Kaba Master Key System provides applications from homes to medium-sized industrial and administrative buildings. (courtesy Kaba Locks, Ltd.)

Fig. 8-43. The Grand Master System provides for several levels of security control. (courtesy Kaba Locks, Ltd.)

single insertion cylinder (Fig. 8-44) has the dimensions shown in Fig. 8-45. The lock is provided with three keys and is registered to protect against unauthorized duplication of the key. The lock is operated from only one side. The lock can be armored, have a knob, cylinder extension, or other special versions.

A double-insertion cylinder (Fig. 8-46) also comes with three keys and is registered. Figure 8-47 shows the dimension specifics. Figure 8-48 shows a double-insertion with a cylinder on one side and a thumbturn on the other, while Fig. 8-49 shows the cylinder with an emergency key function.

This version is particularly suited to schools, hotels, guest houses and senior citizen homes where privacy must be ensured but at the same time an overriding release must be available in case of extreme emergency.

The operation of this lock is as follows: If a key is left inserted and turned on the inside of the door, normal keys cannot open the door from the outside. The advantage of the Kaba 20 double cylinder with the emergency key device means that with the special emergency key, the door can still be opened from the outside, providing entry in case of an emergency. The emergency key device also

fulfills a function required particularly in hotels, whereby inserting a reserve key from the inside and locking the door from the outside, a room is made inaccessible for the guest key as also for the pass key. The emergency key function can only be built in on one side of the lock cylinder.

For exterior and interior doors opening inward (and

Fig. 8-44. Single insertion (one side of lock) cylinder. (courtesy Kaba Locks, Ltd.)

107

Fig. 8-45. Cylinder technical specifications. (courtesy Kaba Locks, Ltd.)

Fig. 8-47. Specifications for a Kaba 20 double cylinder lock. (courtesy Kaba Locks, Ltd.)

even for glass panelled doors), the automatic deadlatch is an excellent choice (Fig. 8-50). The case and staple are precision die-cast. Note that the bolt is not reversible, so the unit is not suitable for any door opening outwards. Dimensions for the narrow stile and standard stile models are at Fig. 8-51.

The lock is operated by the key from the outside and by a turn handle from the inside. The turn handle can also have a key cylinder if desired. The springbolt is automatically deadlocked against end pressure when the door is closed. The springbolt can be held back by the slide button. Please note, though, that the slide button cannot

Fig. 8-46. Kaba double cylinder for access control from either side of door. (courtesy Kaba Locks, Ltd.)

Fig. 8-48. Double cylinder with a thumbturn on one side. (courtesy Kaba Locks, Ltd.)

Fig. 8-49. A double cylinder with an emergency access key. (courtesy Kaba Locks, Ltd.)

be used for locking the bolt in the locked position. The inside handle can be deadlocked by a complete *reverse* turn of the key after closing the door.

The rim deadlock (Fig. 8-52) is most suitable for use with apartment doors, offices, etc. The unit is reversible for either left or right-handed doors. Figure 8-53 provides dimensions for the deadlock.

The rim latch (Fig. 8-54) is suitable for apartment doors or offices, usually above a mortise deadlock. Line the deadlock, an inside turn knob is provided for easy opening from the interior. Dimensions are at Fig. 8-55.

For exterior doors, the mortise deadlock (Fig. 8-56) is an excellent choice and provides a fantastic amount of security not expected of such a lock. This high security mortise deadlock was built to meet the specifications of a thief resistant lock as defined in the British standards. It applies equally well the standards to American thief resistant locks. The dimensions are at Fig. 8-57. The lock, suitable for exterior doors, has an impressive 22mm long bolt throw and is particularly useful for double door installations. The cylinder can be single or double, key on one side and a knob on the other, armored or not.

Cylinder guards and their dimensions are shown in Fig. 8-58. The cylinder guard (center of Fig. 8-58) shows both the exterior and interior sides held together by the cylinder fixing screw.

A heavy-duty two bolt mortise lock and security mortise deadlock are shown at Fig. 8-59. Exchangeable insertion cylinders are available, so that a single, double,

Fig. 8-50. Automatic deadlatch lock. (courtesy Kaba Locks, Ltd.)

Case	84 x 88 x 30mm *3.5/16" x 3.7/16" x 1.3/16"*	64 x 88 x 30mm *2½" x 3.7/16" x 1.3/16"*
Backset	60mm *2¾"*	40mm *1.9/16"*
Springbolt throw	19mm *¾"*	14.25mm *9/16"*
For doors of a thickness from	29 to 57mm *1.1/8" to 2¼"*	29 to 57mm *1.1/8" to 2¼"*

Fig. 8-51. Dimensions for the automatic deadlatch, for standard and narrow stile doors. (courtesy Kaba Locks, Ltd.)

Fig. 8-52. Rim deadlock suitable for either hand of a door. (courtesy Kaba Locks, Ltd.)

Fig. 8-54. Rim latch lock unit. (courtesy Kaba Locks, Ltd.)

or single with a thumbturn on the opposing side can be used in the lock unit. Dimensions are shown at Fig. 8-60.

Security locks for hollow metal doors, as available from Adams Rite, are available and of great use. Figure 8-61 shows several such locks. You will notice that the third lock (with the hookbolt) has a standard Kaba-20

mortise cylinder. A pair of these are supplied, providing locking operation from both sides (Fig. 8-62). The key integrity is provided by the knowledge that there are over 100 million possible key combinations available.

Deadbolt Locks. The Kaba-20 deadbolt lock (Fig. 8-63) comes with an exclusive "inner sanctum core"

Fig. 8-53. Specifications for the rim deadlock. (courtesy Kaba Locks, Ltd.)

110

Fig. 8-55. Specification data for the rim latch lock. (courtesy Kaba Locks, Ltd.)

which bolts onto itself; it cannot be pried apart. The double cylinder can come with a 60mm (2⅜″) or 70mm (2¾″) backset, so it is most applicable to US lockset replacement standards. The deadbolt features the screw-on mortise cylinders in a heavy-duty housing and a deadbolt with an impressive one inch throw and two hardened steel points to prevent the bolt from being cut. The unit is ideal for both new doors or as an auxiliary lock to upgrade the security of an existing door. The double cylinder deadbolt has no exposed screws; the unit is held in place by cylinder set screws under the armored front. Dimensions for the two backset models are shown at Fig. 8-64. The unit also comes with a single cylinder and a turn knob (Fig. 8-65).

Since all Kaba units, like other locks, comes with installation instructions in the package, only one example will be provided here.

Installing the Double Cylinder Deadbolt. Using the template that comes with the lock unit, mark the door as shown in Fig. 8-66. Bore two holes—a 2⅛″ hole through both sides of the door, and a 1″ diameter hole on the door edge. Ensure the holes are perpendicular, and also that the door edge hole is bored last (Fig. 8-67).

Mortise the front. Line up the front with the hole and mark its outline. Mortise along the lines to the proper depth (Fig. 8-68).

Install the cylinder housing assembly into the door. Insert the bolt housing into the 1″ hole and through the cylinder housing. Finally, screw in the two cylinder housing screws (Fig. 8-69).

Assemble the cylinder and wave washer into the cylinder rose as in Fig. 8-70. Note: Include the cylinder ring for 1⅜″ thick doors only! Insert the key into the

cylinder. Pull back the key *one* notch and screw the cylinder into the housing until it stops. You can use the key as a "wrench" in this instance. Now reverse the

Fig. 8-56. The mortise deadlock provides exceptionally high exterior door lock security. (courtesy Kaba Locks, Ltd.)

111

FOR 1¾" (45 mm) DOOR THICKNESS:
Length of Standard Cylinder   65mm
Height of Cylinder Guard   10mm
Alternative dimensions are available.

Fig. 8-57. Mortise deadlock dimension data. (courtesy Kaba Locks, Ltd.)

Fig. 8-58. Cylinder guards for locks and their specifications. (courtesy Kaba Locks, Ltd.)

Fig. 8-59. The Kaba two-bolt mortise (left) and security mortise (right) deadlocks. (courtesy Kaba Locks, Ltd.)

Fig. 8-60. Specification data for the two-bolt and security mortise deadlocks. (courtesy Kaba Locks, Ltd.).

Fig. 8-61. The Kaba locks fit all Adams Rite narrow stile metal door units. (courtesy Kaba Locks, Ltd.)

rotation of the cylinder until the Kaba name is up. Hold the cylinder, push in the key, and turn it to operate the bolt (Fig. 8-71).

Assemble the other cylinder and cylinder rose. Screw into the cylinder housing on the other side until it is tight to the door and the Kaba name is in the up position (Fig. 8-72). Make sure the cylinders line up on both sides of the door. Insert the Allen wrench inside along each side of the bolt until it engages the screw, then tighten the two cylinder set screws (Fig. 8-73).

Attach the front with the two screws provided (Fig. 8-74) and then try the key on both sides and operate the bolt.

Install the box and strike. Line up the strike with the bolt and mark the outline of the strike. Mortise the lines to the depth required. Next, drill two 1″ diameter holes 1″ deep into the jam. Chisel out the holes square for the box and attach, installing the strike over the box (Fig. 8-75).

Figure 8-76 shows an exploded view of the lock unit.

THE CYLINDER KEY

Figure 8-77 illustrates basic key nomenclature, while Fig. 8-78 gives some indication of the variety of blanks available. Notice the differences in the bow, blade length, and in the width, number, and spacing of the grooves.

The *bow* usually has a specific shape that identifies the lock maker. Traditionally, key blank bows were unmarked; today many of these bows mimic the original design of the key. The blade length is indicative of the number of pins, e.g., a five-pin key is shorter than a six-pin key, and a seven-pin key of the same make is longer than either.

The height of the blade implies something about the depth of the cuts in the series; the higher the blade the deeper the cuts.

Key blanks are made of brass, nickel-brass, nickel-steel, steel, aluminum, or a combination of these and other alloys. The metal determines the strength of the key and its resistance to wear.

Key Board

Earlier, I mentioned the need for a key board. The same board can be expanded to include cylinder keys. Key-blank interchange charts can be purchased from the local locksmith supply house. These charts are sometimes published in house catalogs. You might also consider labeling each blank.

Selection

Determine if you have the proper key blank with reference to the manufacturer of the lock. Suppose you need a No. 14 Yale blank and do not have it. Star Key Company's interchange chart shows that Star blank No. 5YA11 will substitute. This blank will fit the lock with the same precision as the original Yale product.

Sometimes you will not know the key-blank number. In this case, examine the bow of the key—its shape is a clue. Look for other blanks that have approximately the same shape. Once you have found several candidates that might be acceptable, compare the key blades. The key section (the shape of the blade when viewed from the front), the blade length, and shoulder configuration should be identical.

Duplicating by Hand

Duplicating a pin-tumbler key by hand is not much different than duplicating a lever or warded key. The prime requirement is patience.

1. Smoke the original and mount it together with the blank in a vise. The blank must be in perfect alignment with the original.

2. Using Swiss round and warding files, begin at the tip of the key and work towards the bow. Start in the center of each cut.

3. Go slow near the end of each cut, but continue to cut with firm and steady strokes. Stop when the file touches the black on the original key.

4. Once you have finished, remove the keys from the vise. Wire-brush the duplicate to remove any burrs and polish the key.

Fig. 8-62. Kaba standard mortise cylinders can also be used on Adams Rite units. (courtesy Kaba Locks, Ltd.)

Fig. 8-63. Double-cylinder deadbolt lock with "Inner Sanctum" core. (courtesy Kaba Locks, Ltd.)

Fig. 8-64. Double cylinder deadbolt specification data. (courtesy Kaba Locks, Ltd.)

Fig. 8-65. Deadbolt also comes in a single cylinder configuration with a thumbturn on the interior side. (courtesy Kaba Locks, Ltd.)

5. Hold the duplicate up to the light and place the original in front of it and then behind it. Is the duplicate a faithful copy of the original? Shallow cuts can be dressed out with a file; cuts that are too deep mean that you will have to start over with a new blank. In extreme cases, the original may be damaged.

Making a substitute for the original is tricky. Use the original as a model, but do not file to the bottom of the grooves. When you have cut as far as you dare, check your work in the lock. File as needed, but with circumspection. Be content to remove some metal and test again. When the key turns, stop.

Fig. 8-66. Mark the holes for drilling. (courtesy Kaba Locks, Ltd.)

Fig. 8-67. Bore the door edge hole last; this ensures the proper alignment between both holes for lock insertion. (courtesy Kaba Locks, Ltd.)

117

Fig. 8-68. Mortise the door edge only to the required depth, no deeper. (courtesy Kaba Locks, Ltd.)

Fig. 8-71. Insert the cylinder as far as possible, then back off until the alignment is proper. (courtesy Kaba Locks, Ltd.)

Fig. 8-69. Insert the cylinder housing before the bolt housing assembly. (courtesy Kaba Locks, Ltd.)

Fig. 8-72. Make sure the "Kaba" name is centered on top of the cylinder. (courtesy Kaba Locks, Ltd.)

Fig. 8-70. Prepare the pin cylinder for insertion, remembering the cylinder wave spring is for 1⅞" door thicknesses only. (courtesy Kaba Locks, Ltd.)

Fig. 8-73. Using the Allen wrench provided to tighten the interior located screws. (courtesy Kaba Locks, Ltd.)

Now you have a working duplicate key and you can leave it at that. But a professional will take time to disassemble the lock and check the pin heights. He may replace a tumbler or modify the key.

Duplicating by Machine

A key-duplicating machine deserves care and respect since it is vital to your livelihood (Fig. 8-79). The machine's instruction sheet details service procedures

Fig. 8-74. Attach the faceplate with the two screws provided. (courtesy Kaba Locks, Ltd.)

Fig. 8-75. Align the strike with the lock faceplate, mark it, then drill the holes and mount the unit. (courtesy Kaba Locks, Ltd.)

and schedules. Once the machine is aligned and the cutter tolerances are in order, proceed as follows:

1. Place the original in one vise and the blank in the vise next to the cutter wheel. The keys should rest on ledges at the base of each vise.

2. Check the tip and the shoulder alignment with the cutter and cutter guide.

3. Put on eye goggles (particles thrown off by the cutter can blind you) and start the machine.

4. Move the blank into the cutter. The guide regulates the cuts so that the blank takes on the same profile as the original. "Rough out" the cuts on the first few passes.

5. Finish the job with one or two light passes. Be thorough: a less-than-perfect duplicate will not open the lock.

6. Wire-brush away any burrs and polish the key. Test it in the lock. It should work as smoothly as the

Fig. 8-76. Exploded view of the Kaba 20 double cylinder deadbolt lock cylinder shows proper parts alignment and positioning. (courtesy Kaba Locks, Ltd.)

Fig. 8-77. Key nomenclature.

original. If not, compare the original and duplicate under a light. File off any rough spots.

Impressioning

The technique for impressioning cylinder keys is different than for warded keys. The cylinder key is not smoked since inserting the key would wipe the black off the top of the blade. Instead, one must depend upon marks left on the blank by the tumblers. These marks indicate the position of the tumblers relative to the blank and the depth of each cut. Proceed as follows:

1. Examine the blank. The marks left by the tumblers are difficult to see in the best of circumstances; scratches across the top of the blade will make them impossible to see. Polish the top of the blade with emery paper.

2. Place the bow of the key in a small C-clamp and insert the blade into the keyway.

3. Twist the blank hard. If done correctly, the tumblers will leave their "footprints" on the blade.

4. Examine the blank under a light. Tip it slightly to give the marks definition.

5. Identify each mark with a light file cut.

6. Make a shallow cut at the mark closest to the tip with a Swiss No. 4 file.

7. Insert the key, twist it back and forth, and remove it. Check the mark and file a little more. Continue until the tumbler no longer leaves a mark. When this happens the pin is resting on the shear line.

Fig. 8-78. A collection of Master blanks.

120

Fig. 8-79. Two key-duplicating machines.

8. When you finish one cut, move to the next and repeat the procedure. Keep the slopes even and smooth.

9. When the last cut is finished, all pins should be at the shear line.

10. Remove the key and clean it. This operation includes smoothing the various angles and removing burrs.

Rock the blank up and down as well as from side to side. It may be useful to work the key with the help of a steel rod. Insert the rod through the hole in the bow of the key and tap the bow with a mallet.

An impressioning tool can be purchased from a locksmith supply house. Squeezing the handle impressions the key.

Chapter 9

Master Keying

Master keying can provide immediate and long range benefits that a beginning locksmith can find most desirable. This chapter covers the principles of master keying, as well as techniques for developing master key systems.

CODING SYSTEMS

Coding systems help the locksmith to distinguish various key cuts and tumbler arrangements. Without coding systems, master keying would be nearly impossible.

Most coding systems (those for disc-, pin-, and lever-tumbler locks) are based upon depth differentiation. Each key cut is coded according to its depth; likewise, each matching tumbler receives the same code. Depths for key cuts and tumblers are standardized for two reasons: (1) It is more economical to standardize these depths. Mass production would be impossible without some kind of standardization. (2) Depths, to some extent, are determined by production.

MASTER KEY SYSTEMS

In most key coding systems, tumblers can be set to any of five possible depths. These depths are usually numbered consecutively 1 through 5. Since most locks have five tumblers, each one having five possible settings, there can be thousands of combinations. Master keys are possible because a single key can be cut to match several lock combinations.

In developing codes, there are certain undesirable combinations which cannot be used. The variation in depths between adjoining tumblers cannot be too great. For example, a pin-tumbler key cannot be cut to the combination 21919 because the cuts for the 9s would rule out the cuts for the 1s. Likewise a pin-tumbler lock with the combination 99999 would be too easy to pick. The undesirable code combinations vary depending upon the type of tumblers, the coding system, and the number of possible key variations. The more complex the system, the greater the possibility of undesirable combinations.

Master Keying Warded Locks

Since a ward is an obstruction within a lock that keeps out certain keys not designed for the lock, a master key for warded locks must be capable of bypassing the wards. Figure 9-1 shows a variety of side ward cuts that are possible on warded keys. The master key (marked *M*) is cut to bypass all the wards in a lock admitting the other six keys.

Fig. 9-1. The master key at the bottom of the drawing replaces the six change keys above it.

As explained earlier, cuts are also made along the length of the bit of a warded key. These cuts correspond to wards in the lock. To bypass such wards, a master key must be narrowed.

Because of the limited spaces on a warded key, master keying is limited in the warded lock. The warded lock, because it offers only a very limited degree of security, uses only the simplest of master key systems. Figure 9-2 shows some of the standard master keys that are available from factories.

Master Keying Lever Locks

Individual lever locks may be master keyed locally, but any system that requires a wide division of keys would have to be set up at the factory. A large selection of tumblers is required. The time involved in assembling a large number would make the job prohibitive for the average locksmith.

There are occasions when you will be asked to master key small lever locks. There are two systems. The first is the double-gate system (Fig. 9-3); the other, the wide-gate system (Fig. 9-4). Double gating is insecure. As the number of gates in the system increases, care must be taken to prevent cross-operation between the

Fig. 9-2. Factory-supplied warded master keys.

Fig. 9-3. Double-gated lever cuts (shown by the dotted lines) should be avoided.

change keys. For example, you may find a change key for one lock acting as the master key.

With either system, begin by determining the tumbler variations for the lock series in question. If the keys to all the locks are available, read the numbers stamped on the keys. Otherwise the locks must be disassembled so that you can note the tumbler depths for each one. Next, make a chart listing the tumbler variations (Fig. 9-5).

The master key combination can be setup fairly easily now. The chart in Fig. 9-5 is for ten lever locks, each having five levers with five possible key depths per lever. The master key for these locks will have a 21244 cutting code.

Suppose the tumblers in the first position have depths of 1, 2, 3, and 5. Depths 1 and 3 must be filed wider to allow the depth 2 cut of the master key to enter. The tumbler with the depth 5 cut requires a separate cut, or double gating. It will have a cut that will align it properly at two positions.

Moving to position 2 on the chart, we see that four levers will have to be cut; all of these will require a double gating cut. In position 3, four will require a double gating and one will need widening. In position 5, only one will require another gating to be cut, while four will require widening at the current gate.

Another master keying method is to have what is known as a master-tumbler lever in each lever lock. The master tumbler has a small peg fixed to it that passes through a slot in the series tumblers. The master key raises the master tumbler. The peg, in turn, raises the individual change tumblers to the proper height so that the bolt post passes through the gate in each lever. The lock is open.

This system should be ordered from the manufacturer. The complexities building one yourself requires superhuman skill and patience.

Master Keying Disc-Tumbler Locks

Disc-tumbler locks have as few as three discs and as many as twelve. The most popular locks have five.

Figure 9-6 shows how the tumbler is modified for master keying. The left side of the tumbler is cut out for the master; the right side responds to the change key. The key used for the master is distinct from the change key in that its design configuration is reversed. The cuts are, of course, different. The keyway in the plug face is patterned to accept both keys. Since the individual disc tumblers are numbered from 1 to 5 according to their depths, it is easy to think of the master key and disc cuts on a 1 to 5 scale but on different planes for both the key and the tumblers.

Uniform cuts are taboo. The series 11111 or 22222 would give very little security since a piece of wire could serve as the key. Other uniform cuts are out because they are susceptible to shimming. To keep the system secure, it is best to keep a two-depth interval between any two change keys. For example, 11134 is only one depth away from 11133, so 11134 should be used and 11133 omitted. The rationale is that 11134 is the more complex of the two.

The next step is to select a master key combination composed of odd numbers. At the top of your worksheet, mark the combination you have selected for the master key. Below it add a random list of possible change-key numbers. If you choose a systematic approach in developing change-key numbers, you compromise security. On the other hand, the systematic approach insures a

Fig. 9-4. Wide-gated cuts are preferred.

TUMBLER POSITIONS

TUMBLER DEPTHS	POSITION 1	POSITION 2	POSITION 3	POSITION 4	POSITION 5
DEPTH 1	4	4		4	1
DEPTH 2	3		5		
DEPTH 3	1	3	1		
DEPTH 4		1	4	3	5
DEPTH 5	2	2		3	4
TOTAL FOR EACH POSITION	10	10	10	10	10

MASTER KEY: 21244

Fig. 9-5. Master keying by the numbers.

complete list of possible combinations. You could begin systematically, then randomly select the change-keys combinations.

A single code could be used for all customers. The main point to remember is to use different keyways.

Master Keying Pin-Tumbler Locks

Master keying is more involved than modifying the cylinder. It requires the addition of another pin sandwiched between a top and bottom pin. This pin is, logically enough, called the *master pin*.

The master system is limited only by the cuts allowed on a key, the number of pins, and the number of pin depths available. Since this book is for beginning and advanced students, I will cover the subject on two levels: the simple master key system for no more than 40 locks in

a series, and the more complex system involving more than 200 individual locks.

It is important to remember that a master key system should be designed in such a way as to prevent accidental cross keying.

Pins are selected on the basis of their diameters and lengths. Master pin lengths are built around the differences between the individual pin lengths. Consider, for a moment, the Yale five-pin cylinder, with pin lengths ranging from 0 to 9 cuts. Each pin is 0.115 in. in diameter. Lower pin lengths are:

0 = 0.184″		5 = 0.276″	
1 = 0.203″		6 = 0.296″	
2 = 0.221″		7 = 0.315″	
3 = 0.240″		8 = 0.334″	
4 = 0.258″		9 = 0.393″	

Fig. 9-6. The master key operates on the left side of the tumbler.

MASTER KEY INSERTED

Fig. 9-7. Master keying pin-tumbler locks means a combination other than the change key will raise the pins to the shear line.

As an illustration, let's master key ten locks with five tumblers each. Each tumbler can have any of ten different individual depths in the chamber.

1. Determine the lengths of each pin in each cylinder. Mark these down on your worksheet. The master key selected for this system may have one or more cuts identical to the change keys.

2. Cut a master key to the required depths. In this instance, each cylinder plug is loaded by hand.

3. Using the known master key depth, subtract the depth of the change key from it. The difference is the length of the master key pin. If the change key is 46794 and the master key is 68495, the master pin combination will be as follows:

CHANGE KEY INSERTED

GRAND MASTER INSERTED

MASTER KEY INSERTED

GREAT GRAND MASTER INSERTED

Fig. 9-8. The Yale great grand master system requires five master pins.

Chamber Position	Bottom	Master
1	4	2
2	6	2
3	7	3
4	9	0
5	4	1

This procedure is followed for each plug. Notice that not all the chambers have a master pin. Such complexity is not necessary and makes the lock more vulnerable to picking. Each master pin represents another opportunity to align the pins with the shear line. Figure 9-7 is an extreme instance, with five master pins and three master keys.

In practice, locksmiths avoid most of this arithmetic by compensating as they go along. For example, an unmaster keyed cylinder has double pin sets. Master keying means that an additional pin is added in (usually) position 1. This drives the bottom pin lower into the keyway. The bottom pin must be shortened to compensate.

A grand master key or a great grand master key adds complexity to the system. You have two choices: You can add master pins in adjacent chambers (the Yale approach shown in Fig. 9-8) or you may stack pins in the first chamber. Suppose you have a No. 4 bottom pin and the appropriate master pin (Fig. 9-9A). In order to use grand and great grand master keys, you must add two No. 2 pins so that all four pins will operate the lock (Fig. 9-9B).

DEVELOPING THE MASTER KEY SYSTEM

A master key system should be planned to give the customer the best security that the hardware is capable of. Begin by asking the customer these questions:

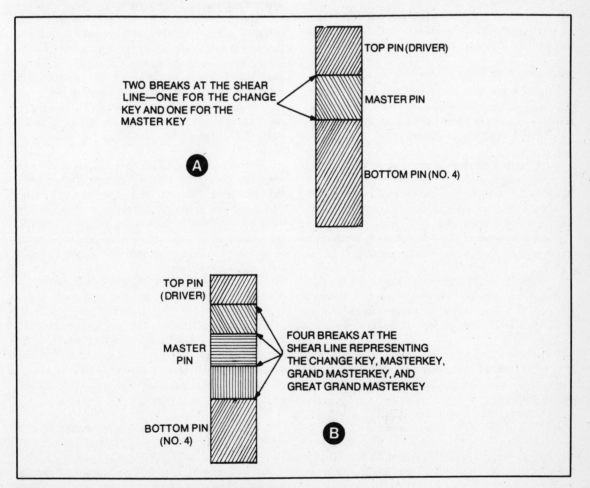

Fig. 9-9. Stacking pins is an alternative way of master keying.

- [ ] Do you want a straight master key system with one master to open all locks? Or do you want a system that will have submasters? That is, do you want a system with several submasters of limited utility and a grand or great grand master?
- [ ] What type of organizational structure is within the business? Who should have access to the various levels?
- [ ] How many locks will be in each submaster grouping?
- [ ] Is the system to be integrated into an existing system, or will the system be developed from scratch?
- [ ] What type of locks do you have?

Once you have (with the help of your catalogs, specification sheets, technical bulletins, and experience) digested these answers, you are ready to begin development of a master key system.

The purpose of master keying is to control access. A key may open one lock, a series of locks, a group composed of two or more series, or every lock in the system.

Master Key Terms

Before we go much further into the subject, it is wise to spend a few minutes defining the terms. This glossary has been prepared by Eaton Corporation (Yale locks) and is reprinted with their permission.

bicentric cylinder—A pin-tumbler cylinder with two plugs, which effectively make it two cylinders in one. The bicentric cylinder is recommended for large, multilevel master key systems, where maximum security and expansion is required.

building master key—A master key which opens all or most of the locks in an entire building.

change key (or individual key)—A key which will usually operate only one lock in a series, as distinguished from a master key which will operate all locks in a series. Change keys are the lowest level in a master key system.

changes (key)—See **key changes**.

construction breakout key (CBOK)—A key used by the owner to make all construction master keys permanently inoperative.

construction master key (CMK)—A key normally used by the builder's personnel for a temporary period during construction. It operates all cylinders designated for its use. The key is permanently voided by the owner when he accepts the building or buildings from the contractor.

control key—A key used to remove the central core from a removable core cylinder.

controlled cross keying—See **cross keying**.

cross keying—When two or more different change keys (usually in a master key system) intentionally operates the same lock.

cross keying controlled—When two or more change keys under the same master key operate one cylinder.

cross keying uncontrolled—When two or more change keys under *different* master keys operate one cylinder.

department master key—A master key that gives access to all areas under the jurisdiction of a particular department in an organization, regardless of where these areas are in a building or group of buildings.

display room key—A special hotel change key that will allow access to only designated, even if the lock is in the shutout mode. With many types of hotel locks, this key will also act as a shutout key, making all other change and master keys inoperative, except the appropriate individual display room key and the emergency key.

dummy cylinder—One without an operating mechanism; used to improve the appearance of certain types of installations.

emergency key (EMK)—A special, usually top level, hotel master key that will operate all the locks in the hotel at all times. An emergency key will open a guestroom lock even if it is in the shutout mode. With many types of hotel locks, this key will act as a shutout key, making all other change and master keys inoperative, except the appropriate individual display room key and the emergency key.

engineer's key (ENG)—A selective master key which is used by various maintenance personnel to gain access through many doors under different master and grand master keys. The key can be set to operate any lock in a master key system and, typically, fits building entrances, corridors, and mechanical spaces. Establishing such a key avoids issuing high level master keys to maintenance personnel. See also **selective master key**.

floor master key—A master key that opens all or most of the locks on a particular floor of a building.

grand master key (GM or GMK)—A key that operates a large number of keyed-different or keyed-alike locks. Each lock is usually provided with its own change key. The locks are divided into two or more groups, each operated by a different master key. Each group can be operated by a master key only, or by the grand only, or by a master and the grand.

great grand master key (GGM or GGMK)—A key that operates a large number of keyed-different and/or keyed-alike locks. Each lock is usually provided with its own change key. The locks are divided into two or more groups, each operated by a different grand master key. Each of these groups is further subdivided into two or more groups, each operated by a different master key. A group can be operated by the great grand only, or a grand only, or a master only, or any combination of the three.

guestroom change key—The hotel room key that is normally issued to open only the one room for which it was intended. The guestroom key cannot be used to set a hotel lock in the shutout mode from the outside of the room, nor will it open a hotel lock from the outside if the lock is in the shutout mode.

hotel-function shutout—When a hotel-function lock is in the shutout mode, regular master keys of all levels and the guestroom key will not open the lock from the outside of the room. Most hotel-function locks can be set in the shutout mode with the thumbturn or pushbutton from the inside, or with the emergency or display room key from the outside.

hotel keying—Keying for hotel-function locks.

hotel great grand master keys, grand master keys, or master keys—Depending on the level of the system, these keys function as they would in a normal system except they cannot be used to set a hotel lock in the shutout mode from the outside of the room, nor can they open a hotel lock from the outside if the lock is in the shutout mode.

housekeeper's key—A grand master key in a great grand level hotel system that normally operates all the guestrooms and linen closets in the hotel.

interchange—See **key interchange**.

key bitting—A number that represents the depth of a cut on a tumbler-type key. A bitting is often expressed as a series of numbers or letters that designate the cuts on a key. The bittings on a key are the cuts which actually mate with the tumblers in the lock.

key bitting depth—The depth of a cut that is made in the blade of a key. See also **root depth**.

key bitting list—A list originated and updated by the lock manufacturer for every master key system established. This list contains the key bittings of every master key and change key used in the system. Each time an addition is made to the system, all new bittings used are added to the list. It is essential that a complete copy of this list be furnished to any personnel servicing a master key system locally. The lock manufacturer should be informed of any changes made locally to a keying system.

key bitting or cut position (also called spacing)—The location of each cut along the length of a key blade. It is determined by the location of each tumbler in the lock. Bitting position is measured from a reference point to the center of each cut on the key. The most common reference point is the key stop, but the tip of the key is sometimes used.

key change number—A recorded number, usually stamped on the key for identification. A key change number can be either the direct bitting on the key or a code number.

key changes (chges)—The total possible number of different keys available for a given type of tumbler mechanism. In master key work, the number of different change keys available in a given master key system.

key interchange—An undesirable situation, usually in a master key system, whereby the change key for one lock unintentionally fits other locks in the system.

key section (KS)—The cross-sectional shape of a key blade that can restrict its insertion into the lock mechanism through the keyway. Each key section is assigned a designation or code by the manufacturer. A key section is usually shown as a cross section viewed from the bow towards the tip of the key. See also **keyway**.

key set (or set)—A group of locks keyed exactly the same way. A key set is usually identified with a key symbol. See also **standard key symbols**.

keyed alike (KA)—A group of locks operated by the same change key. Not to be confused with master keying.

keyed different (KD)—A group of locks each operated by a different change key.

keying—A term used in the hardware industry that refers to the arrangement of locks and keys into groups in order to limit access.

keying levels—The stratification of a master key system into hierarchies of access. Keying systems are available with one or more levels. The degree of complexity of the system depends on the number of levels used. Generally, the top level master key can open all locks in the system. Each successive intermediate level of master key can open fewer locks but can open more locks than a change key.

keying system chart—A chart indicating the structure and expansion of a master key system, showing the key symbol and function of every master key of every level.

keyway (Kwy)—The shape of the hole in the lock mechanism that allows only a key with the proper key section to enter. See also **key section**.

levels—See **keying levels**.

maid's key—A hotel master key given to the maid which will give access only to the guestrooms and linen closets in her designated area of responsibility. A hotel is normally divided into floors or sections with a different maid's key for each floor or section. A maid's key will not open a guestroom if the lock is in the shutout mode.

master key (MK or Mky)—A key that operates a series of locks, each of which has its own change key.

master keyed lock—A lock designed so it can be operated by its own change key, as well as by a master key.

multiple key section system—(also called **sectional key sections**)—Used to expand a master key system by repeating the same or similar key bittings on different key sections. Keys of one section will not enter locks with a different section, yet there is a master key section milled so it will enter some or all of the different keyways in the system. See also: **simplex key section**.

paracentric keyway—A keyway in a cylinder lock with one or more side wards on each side projecting beyond the vertical centerline of the keyway to hinder picking. See also **simplex key section**.

pin tumblers—Small sliding pins in a lock cylinder, working against drivers and springs and preventing the cylinder plug from rotating until raised to the exact height by the bitting of a key.

plug (of a lock cylinder)—The round part containing the keyway and rotated by the key to transmit motion to the bolt or other locking mechanism.

plug retainer—The part of a lock cylinder which holds the plug in the shell.

privacy key—A change key set up as part of a master key system but not operated by any master keys or grand masters of any level. This key is set up for such areas as liquor-storage rooms in hotels, narcotic cabinets in hospitals, and food storage closets where valuables are kept.

removable core cylinder—A cylinder containing an easily removable assembly which holds the entire tumbler mechanism, including the plug, tumblers, and separate shell. The cores are removable and interchangeable with other types of locks of a given manufacturer by use of a special key called the control key. See also **control key**.

root depth—Refers to the distance from the bottom of a cut on a key down to the base or bottom of the key blade. Root depth is easy to determine since it measures the amount of blade remaining, rather than the amount which was cut away (bitting depth).

selective master key—A special top level master key in a grand or great grand system that can be set to operate any lock in the entire system, in addition to the regular floor or section master key, without cross keying. Typical selective master keys include an engineer's key (ENG), nurse's key (NUR), and attendant's key (ATT). The number of selective master keys is normally limited to one or two and should be setup when the original system is established. See **engineer's key**.

simplex key section—A single independent key section that cannot be expanded into a multiple key section system. Simplex key sections such as the Yale "Para" are used for stock locks and small master key systems.

spacing—See **bitting position**.

standard key symbols—A uniform way of designating all keys and cylinders in a master key system. The symbol automatically indicates the exact function of each key or cylinder in the system, without further explanation.

tailpiece—The connecting link attached to the end of a rim cylinder which transmits the rotary motion of the key through the door into the locking mechanism.

top level master key—The highest level master key in a multilevel keying system that fits most of the locks in the system.

tumbler—One or more movable obstructions in a lock mechanism which dog or prevent the motion of the bolt or rotation or the plug and that are aligned by the key to remove the obstruction during locking or unlocking.

uncontrolled cross keying—See **cross keying**.

visual key control—A system of stamping all keys and the plug face of all lock cylinders with standard key symbols for identification purposes. Other key and cylinder stamping arrangements are available but are not considered visual key control.

Standard Key Symbol Code

Figure 9-10 illustrates keying levels of control and the rudiments of the standard key symbol code. Great grand master keys are identified by the letters GGM. Grand masters carry a single letter, beginning with the first in the alphabet and identifying the hierarchy of locks that the individual grand master keys open. Master keys carry two letters; the first identifies its grand master, the second identifies the series of locks under it. Thus, a master key labeled AA is in grand master series A and opens locks in master key series A. Master key AB is under the same grand master, but opens locks in series B. Master BA is under grand master B and opens locks in series BA. Change keys are identified by their master key and carry numerical suffixes to show the particular lock that they open. Any key in the series can be traced up and down in the hierarchy. Thus, if you misplace change key AB4, you know that master key AB, grand master A, or the great grand master will open the lock.

There are special keys that are out of series. Some of these keys are mentioned in the glossary, together with the appropriate key symbols for them.

If cross keying is introduced into the system—that is, if a key can open other locks on its level, the cylinder symbol should be prefixed with an X. If the cylinder has its own key, it is identified with the standard suffix. For example, XAA4 is a change-key cylinder that is fourth on this level. It may be cross keyed with AA3 or any other cylinder or cylinders on this level. By the same token, master key AA, grand master A, and the great grand master will open it. Elevator cylinders are often cross keyed without having an individual change key. It is no advantage to have a key that will operate the elevator cylinder and none other in the system. These cylinders are identified as X1X, X2X, and so on.

The symbols that involve cross keying apply to the cylinder only; all other symbols apply to the cylinder *and* the key. This point may seem esoteric, but ignoring it causes the factory and everybody else grief. There is, for example, no such thing as an X1X key. Nor is there an XAA4 key. Change key AA4 fits cylinder XAA4 that

Fig. 9-10. Key levels.

happens to be cross keyed with another cylinder. Key AA4 does not fit any cylinder except XAA4.

There are certain advantages to using the standard key symbol code:

☐ It is a standardized method for setting up the keying systems.
☐ It maintains continuity from one order to the next.
☐ It indicates the position of each key and each cylinder in the hierarchy.
☐ It helps to control cross keying, since each cross keyed cylinder is clearly marked.
☐ It offers a method of projecting future keying requirements.
☐ It can be easily rendered on a chart.
☐ It allows better control of the individual keys within the system.
☐ It is a simple method of selling and explaining the keying system to the architect or building owner.

Selling the System

It is important to be able to communicate the advantages of the key symbol code and the implications it has for setting up an ordered, coherent, and secure keying system. This may take some selling on your part, since architects and building owners tend to think of keys and locks as individual entities and not part of a larger system. Selling a comprehensive master keying system involves the following:

☐ Being able to explain the subject of master keying to the architect or owner.
☐ Reviewing the plans of the building(s).
☐ Choosing the proper level of control required.
☐ Selecting a keying system that will cover present and future expansion requirements.
☐ Presenting the system to the architect or owner at a meeting dealing *specifically* with keying.
☐ Recording all changes to the system as agreed upon at the meeting.
☐ Marking plans with the proper key symbol on the side of each door where the key is to operate.
☐ Presenting the owner with a schematic layout of the entire system, showing him the layout of the masters, grand masters, etc.

MASTER KEY SYSTEM VARIATIONS

So far I have discussed the structure of the master key system. Within this broad structure there are many opportunities for variations. Some of these variations involve the possible range of keyways; others involve

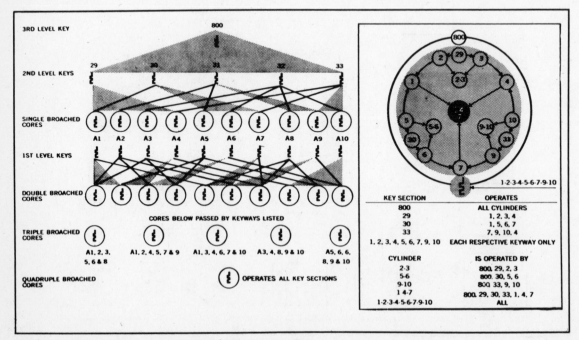

Fig. 9-11. A keyway system developed by the General Lock Company.

Fig. 9-12. General locks have special features described in the text.

special hardware, such as removable cylinders, master-ring cylinders, and rotating tumblers. Each of these variations can extend the range, flexibility, and security of the system. A locksmith must be conversant with all of them.

Keyway Variations

Figure 9-11 illustrates a system of control based on keyway design. General Lock's series 800 key section passes all cylinders in the system; 29, 30, 31, 32, and 33 are submasters, each passing four cylinders; the change keys are restricted to their own individual cylinders. In its fullest expansion, the system includes 35 different keyways on the change-key level and 4 master key levels.

High Security Pin-Tumbler Cylinders

Figure 9-12 illustrates features of the General lock:

☐ A 100° key bitting for long wear (A).
☐ Master pins have a minimum length of 0.040 in. (B). Shorter pins tend to wedge in the chamber.
☐ Only two pins are standard (C).

Fig. 9-13. Corbin's High Security Locking System depends upon split and rotating pins.

Fig. 9-14. Pins are rotated in either direction to allow the joint to uncouple. (courtesy Emhart Corp.)

☐ Pins and springs are made of corrosion-resistant alloy (D and E).

☐ The keyway is part of the security system (F).

☐ If requested, the factory supplies special identification for keys, cores, and cylinders (G).

The Emhart (Corbin) High Security Locking System uses rotating and interlocking pins (Fig. 9-13). The pins must be raised to the shear line and, at the same time, rotated 20° so the coupling can disengage (Figs. 9-14 and 9-15). Rotation is by virtue of the skew-cut bitting on the key (Fig. 9-16). Figure 9-17 illustrates the way the cylinder is armored. The pins are protected by hardened rods and a crescent-shaped shield.

Fig. 9-15. Pins must be rotated and brought to the shear line for the lock to open.

Fig. 9-16. The Corbin key has its bitting cut at 20° angles.

Removable-Core Cylinders

Removable-core cylinders are increasingly popular. Figure 9-18 illustrates the Corbin cylinder, a type that is typical of most. To rekey the change key, follow this procedure:

1. Obtain a Corbin rekeying kit. The kit includes the necessary pins, gauges, and tools.

2. Mount the cylinder in a vise.

3. Remove the plug retainer.

4. Select the key with the deepest bitting as the plug extractor. Normally the grand master key meets this specification; however, there are instances where the engineer's key will have the deepest bitting. A shallow-cut key complicates matters by forcing the control pins, drivers, and buildup pins into the cylinder.

5. Withdraw the plug and remove all pins from their chambers. Figure 9-19 illustrates this procedure.

6. Determine the bitting of the change key.

7. Write down the new combination. As an example, suppose the original change-key bitting is 513525 and we wish to reverse it to 525315 (Fig. 9-20).

8. Install the tumbler pins, ball end down.

9. Use a depth gauge to determine the master key bitting.

10. Calculate the master pins by subtracting the change-key bitting combination from the master key combination. If the master key combination were 525763, the difference between it and the new change-key combination would be 448.

11. Insert the master key into the plug.

12. Install the appropriate master pins (Fig. 9-21).

13. Remove the master key carefully and insert the grand master key. Select the master split pins by subtracting the master key bitting from the grand master key bitting. All pins should be flush with the surface of the plug.

14. Assemble the plug and cylinder.

15. Test all keys.

16. Lubricate the keyway with a pinch of powdered graphite.

Master-Ring Cylinders

The Corbin master-ring cylinder is shown in Fig. 9-22. The change key operates the plug plungers, and the master key operates the plunger in the master ring. Sometimes called "two-in-one" cylinders, these cylinders increase the range of key combinations for any given system. To rekey the change key, follow this procedure:

Fig. 9-17. Passive defense measures include hardened steel pins and armor plate. (courtesy Emhart Corp.)

PROTECTIVE DRILL-RESISTANT SHIELD

DRILL-RESISTANT HARDENED STEEL ARMOR RODS

CYLINDER PLUG HAS SIX CIRCULAR GROOVES, CENTRAL TO TUMBLER PIN HOLES

Fig. 9-18. Corbin removable-core cylinder.

Fig. 9-19. Corbin master-ring cylinder.

GRAND MASTER KEY

FOLLOWER

DUMP ALL PINS FROM THE PLUG

Fig. 9-20. Using the appropriate follower, remove the plug. Dump the pins. (courtesy Emhart Corp.)

1. Mount the plug in a vise and remove the cylinder slide with a pair of pliers or a small chisel.

2. Remove the springs, drivers, and pins (Fig. 9-23).

3. Ream pin holes through the shell, master ring, and plug (Fig. 9-24).

4. Assuming that the original combination was 414472, reversing the combination gives 274414. This will be the combination of the new change key.

5. Reverse the pins to conform with the new combination. That is, the pin that was first goes into the last chamber; the pin that was second goes in to the fifth chamber, and so on.

6. As each pin is installed, tamp it home with a drill bit and turn the key. If the key will not turn, you have confused the pin sequence.

7. Assemble the lock and test.

To rekey the master key, follow this sequence:

1. Mount the cylinder in a vise and remove the cylinder slide with a pair of pliers or a small chisel.

2. Remove the springs, drivers, and pins.

3. Determine the master key bitting with a gauge. The combination runs from the shoulder to the tip of the key, the reverse of the usual sequence. As an example, let it be 678572.

4. Write down the change-key combination and reverse it. Suppose the combination is 275414. Reversed, it is 414572.

CHANGE KEY   5 2 5 3 1 5

Fig. 9-21. Pin length corresponds to the change-key combination. (courtesy Emhart Corp.)

MASTER KEY   4 4 8

Fig. 9-22. Subtract the change-key combination from the master key combination. The difference represents the length of the master pins. (courtesy Emhart Corp.)

Fig. 9-23. After the slide is withdrawn, remove springs, pins, and drivers. (courtesy Emhart Corp.)

5. Subtract the reversed change-key combination from the master key combination (678572—414572). The difference is 264000.

6. Insert the master key into the cylinder and select the appropriate buildup pins. In this case the pins are 264.

Note: If any number in the change-key combination is greater than the master key number above it, you must use a negative number buildup pin in the chamber. For example, a 678572 master key combination with an 814572 change-key combination requires a —2 buildup pin, together with a 6 and 4 pin.

7. As each buildup pin is installed, seat it with a drill bit and turn the key to determine that the correct pin has been installed.

8. Insert the drivers into the cylinder chambers. Either of two drivers are used, depending upon the lock style. Spool drivers (No. J-172) are furnished with mortise cylinders; straight drivers (No. M-099) are used on other cylinders.

9. Insert the springs into their chambers.

10. Holding the springs down with your thumb, try the change key. Do the same for the master and grand master key.

11. Mount the cylinder in the slide, hammering the slide down for a secure fit. Be careful not to damage the threads on mortise cylinders.

12. Try all the keys.

A SIMPLE MASTER KEY SYSTEM

When it becomes necessary to install a large number of locks for any given concern, be it a business or a large residence, wherein the customer wants the locks master keyed, then it is time that your expertise in the realm of the master key systems is put to the test.

Probably the hardest part of master keying development is the creation of a working master key system that takes into consideration any variables that may be required by the customer. For this reason, many locksmiths do not perform their own master keying of any systems; rather, they rely upon the many years of experience, expertise, and professionalism from a factory-developed system. The use of a factory system ensures that the system will meet all the criteria set forth by the customer, the cylinders will be the right ones, and the possibility of extending the system in the future can be assured. For the locksmith, time, personnel, and money are saved through this process.

On the preceding pages, you have been introduced to various aspects of master keying, the types of locks that can be used, the differences in the systems, and the potential for system expansion at a later date. We will not be detailing the varied intricacies of such a system, but will view a small system that you may want to consider and keep in the back of your mind for those small jobs that come up. These are jobs wherein an existing system may be master keyed, using the same cylinders, or a job where a residence is to be master keyed for a homeowner.

Figure 9-25 is a sample master key system developed for a three-floor office building. In addition to the individual office keys on each floor, you have the master key, extra cylinders for additional inner offices on each floor (or for use as replacement cylinders, a main door key, a key for the building maintenance shop, lavatory keys for both the men's and women's rooms on the three floors, and also the start of an additional key code system you can develop should the building or other facility

Fig. 9-24. Ream the pin holes through the shell, master ring, and plug. (courtesy Emhart Corp.)

MASTER: 431472

First Floor	Second Floor	Third Floor
522534	652534	752534
522544	652544	752544
522554	652554	792554
522564	652564	792564
522574	652574	792574
522594	652584	792584
522504	652594	792594
522524	652504	792504
522634	652624	792624
522644	652634	792634
522654	652644	792644
522664	652654	792654
522674	652664	792664
522684	652674	792674
522694	652684	792684
522604	652694	792694
522624	652734	792604
522724	652744	792734
522734	652754	792744
522744	652764	792754

Extra cylinders for each floor:

552784	652514	752614
552514	652524	752601
552614	604714	752602

Main Door Key: 346572
Bldg Maintenance Shop Key: 434672
Construction Key: 466662

NOTE: If building were to have four or more floors, then the last three
digits of the various individual keys could be as above, but the first three digits would be:

Fourth floor: 514___
Fifth floor: 614___
Sixth floor: 714___

If each floor is to have two lavatories, they could be keyed as follows:

First floor: 731573   331577
Second floor: 831573   331578
Third floor: 931573   331579

Fig. 9-25. Sample master key system for a three-floor building.

require more individual keys than the system was designed for.

Within this system, depending upon the size of the installation you will be developing, select the number of keys for individual locks to be used. In this case we are talking about 20 individual door locks on each floor. The master key has already been designated. The progression in the development of the master key system is that the depths of the various individual keys are such that any given key can be recut to become a master key.

In actually rekeying the various cylinders, you will need the standard pins for each cylinder, and then a complete set of master pins. In this case, though, you may choose to use the lower ranking pins of the current system. Why? Because the cylinder pins are worn in addition to the interior of the cylinder itself. Also, in this case, it will save you time and money in having to order the pins.

What you are doing is determining for each key the depths that will be required for each pinning. As an example, let's take two cylinders from the first floor for

rekeying. The first cylinder, 522534 will have the following pinning:

Regular Pin: 4 - 2 - 1 - 4 - 3 - 2
Master Pin: 1 - 1 - 1 - 1 - 4 - 2

With these pins in the cylinder, with the driver pins and tension spring included, either the individual or the master key will open the lock.

For the second cylinder in the series, 522534, the pinning is as follows:

Regular Pin: 4 - 2 - 1 - 4 - 4 - 2
Master Pin: 1 - 1 - 1 - 1 - 3 - 2

Again, with these pins, both the individual and the master key will operate the lock.

Now, how were these particular pins determined? The best way to figure this out is to lay out a simple chart,

Fig. 9-26. Sample chart used for arranging individual cylinder pinning for a master key system.

MASTER KEY PINNING CHART

Master Key # 431472
Individual Key # 522534

Fig. 9-27. Layout chart with master key and individual key cylinder pinning indicated.

as shown in Fig. 9-26. As you progress in selecting pins for the individual cylinders, you first mark the appropriate points on the chart that indicate the master key number. (You can preprint this on a form and save a lot of time.) Then, as you select each cylinder to work with, you put down on the chart the code number for that particular key (Fig. 9-27). The difference between the two points, if any, is the difference between the code number variations. This means that the pinning difference must be the same also.

In Fig. 9-27, for column six, you see that the individual key requires a 4 pin, but the master key requires a 2 pin. The difference of 2 is obvious. The maximum allowable height for the overall height of both pins combined must be 4, so you will require a 2 pin for the master key to operate this particular pin, but also a 2 pin for the individual key. If you only had the 2 pin in the lock, only the master key would be able to open the cylinder at this point; by adding the second 2 pin, you make it possible for the individual key to also open the lock.

If you continue onward, you can readily see how each individual pinning is to be accomplished, what pins will be required, and the insertion sequence into the lock cylinder.

141

Chapter 10

Double-Bitted Locks and Keys

The Junkunc Brothers American double-bitted cylinder lock is usually found in padlocks and office and utility locks (Fig. 10-1).

OPERATION

When the double-bitted key is inserted, it passes through the center of the tumblers (as in a disc lock) and aligns them to the shear line, allowing the plug to be rotated. But the key and the tumbler arrangement is different from that of the regular disc lock.

The key cuts are wavy in appearance; thus the tumblers have to align in a wavelike configuration for the lock to open. Further, there are no definitive tumbler cuts on the key. This is because the key holds ten or more tumblers compressed together and held in a locked position by means of a Z-shaped wire within the tumblers.

All the tumblers are uncoded, meaning they are all of a standard cut. So in order for them to turn within the cylinder, the tumblers have to be cut down. A special keying tool is used for this purpose.

CUTTING DOWN THE TUMBLERS

1. Once the tumblers have been inserted into the plug (with the tumbler spring in place), insert the precut key into the plug.

2. Mount the plug in your vise firmly.

3. Attach the keying tool to a ¼″ drill. Drill the back of the plug. Since the inside diameter of the drill is the same as the outside diameter of the plug, the individual tumblers will cut down to what will be the shear line.

4. Trim the tumblers with a light wire brush to take off any burrs.

5. Insert the plug into the cylinder and test it. Attach the retainer screw and withdraw the key.

The tumblers, since they are uncoded, can be used within any plug. The tumbler spring, because of the shape, holds the various tumblers in the locked position. Only with the insertion of a key, which forces the tumblers into another position, can the lock be opened.

KEYS AND KEYWAYS

The double-bitted lock takes four basic key sections (Fig. 10-2). These sections, of course, match the shapes of the keyways. Keyway one is referred to as a K4 and the

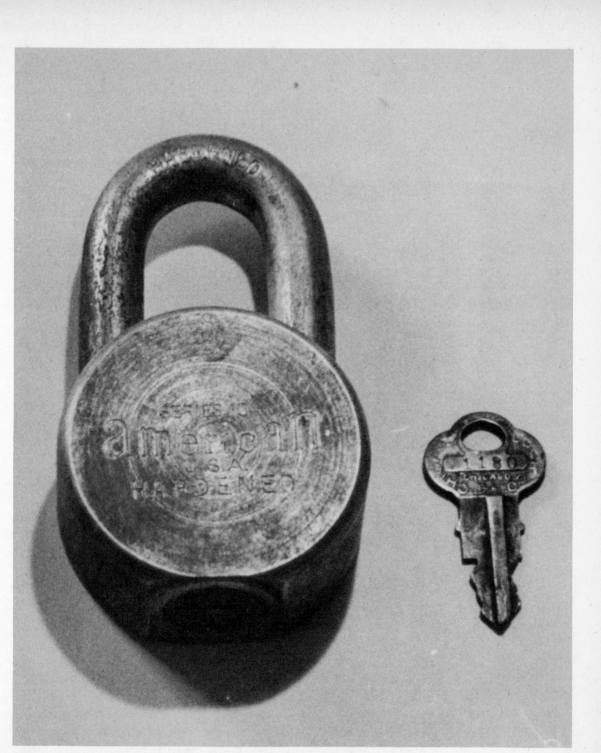

Fig. 10-1. A double-bitted lock and key. (courtesy American Lock Company.)

Fig. 10-2. The four basic double-bitted key sections. (courtesy Taylor Lock Company)

center point is at the center of the tumbler. Keyway two is referred to as a K4L; the center point is just left of center. Keyway three is called a K4R; the center point is right of center. Keyway four, called a K4W, is shaped like a *W*. The keyway shape does not reflect the tumbler types that are within any given plug.

Chapter 11

Vending Machine Locks

The Ace lock, manufactured by the Chicago Lock Company, has become standard on vending machines. The arrangement of the pin tumblers increases security and requires a different type of key—one that is tubular. A typical Ace lock and key is shown in Fig. 11-1. The cam works directly off the end of the plug.

The key has its bitting disposed radially on its end. The depth and spacing of each cut must match the pin arrangement. In addition there are two notches, one on the inside, the other on the outer edge of the key. These notches align the key to the lockface. Otherwise the key could enter at any position.

The key bittings push the pins back, bringing them to the shear line. Once the pins are in alignment, the key is free to turn the plug and attached cam.

The pins within the Ace lock are entirely conventional in construction, with the exception of the bottom one. A ball bearing is sandwiched between the pin and its driver. The bearing reduces friction and increases pin life. The pin in question is not interchangeable with others in the lock. Pin tolerance is extremely critical. There is no room for sloppy key cutting.

DISASSEMBLY

To disassemble the Ace lock, follow this procedure:

1. Place the lock into its holder. Ace makes a special vise for these cylinders.

2. Drill out the retainer pin at the top of the assembly. Use a No. 29 drill bit and stop before the bit bottoms in the hole (Fig. 11-2).

3. Remove what is left of the pin with a screw extractor.

4. Insert the appropriate plug follower into the cylinder. Apply light pressure. Plug follower dimensions are: length 1.50"; outside diameter 0.375"; inside diameter 0.312".

5. Lift off the outer casing from the bushing assembly.

6. Scribe reference marks on the plug sections as an assembly guide.

7. Remove the follower from the cylinder. *Note:* Perform this operation carefully. The pins are under spring tension and must be kept in order. If not, you will have a monumental job sorting the pins. There are 823.543 possible combinations.

ASSEMBLY

Assembly is the reverse order of disassembly. Replace the retaining pin with an Ace part, available from locksmith supply houses.

Fig. 11-1. The Ace lock.

REKEYING

Rekeying is not difficult if you approach the job in an orderly manner.

1. Cut the key. There are seven bit depths, ranging from 0.020″ to 0.110″ in 0.015″ increments.

2. Mount the bushing assembly upright.

3. Remove the pins with tweezers. *Note*: Pin lengths range from 0.025″ to 0.295″ in increments of 0.015″. Drivers are available in 0.125″, 0.140″, and 0.180″ lengths.

4. Select new pins, using the key combination as a guide.

5. Install the pins. The flat ends of the core pins are toward the key (See Fig. 11-1). This pattern must be followed:

Core pins 1, 2, and 3 require 0.180″ drivers.
Core pins 4 and 5 require 0.140″ drivers.
Core pins 6 and 7 require 0.125″ drivers.

Pins are numbered clockwise from the top as you face the lock.

Fig. 11-2. Use a No. 29 drill bit to remove the retaining pin. Alternately, use a No. 42 bit, thread a small metal screw into the hole and extract the pin and screw by prying upward on the screwhead. (courtesy Desert Publications)

PILOT GUIDING ON
CENTER POST

CUTTER REMOVING
INNER LIP OF CASE

Fig. 11-3. Drilling the lock requires a piloted hole saw, available from locksmith supply houses. (courtesy Desert Publications)

6. Insert the key.

7. Install the plug in the cylinder with the scribe marks aligned.

8. Insert a new retaining pin and give it a sharp tap with a punch.

9. Test the key. It may be necessary to rap the cam end of the plug with a mallet.

LOCKOUT

A lockout can be a real headache. Ordinary Ace locks—those without ball bearings—can be drilled out with the tool shown in Fig. 11-3. Hole saws are available for standard and oversized keyways.

Chapter 12

Keyed Padlocks

Padlocks have many uses. They are used to secure out-buildings, bicycles, buildings under construction, tool boxes, paint lockers, and even automobile hoods. Because of this wide use, it pays a locksmith to have a good knowledge of these locks.

While padlock exteriors vary, the functional and operational differences are few and are similar to other locks. Padlocks may use pin tumblers, wards, wafers, levers, or a spring bar. Some must be shackled closed before the key can be removed; this feature is made possible by a spring-loaded coupling. Key security is improved and there is less likelihood of leaving the lock open.

CHOOSING A PADLOCK

Ask the customer if he has a brand preference and then ask:

☐ If width, case length, and shackle clearance are critical (Fig. 12-1).
☐ Where the lock will be used.
☐ How often the lock will be opened. The price of the lock has a direct relationship to its wearing qualities.
☐ If the lock is intended to secure valuable property. An

inexpensive lever lock would be adequate to keep children from straying into the backyard, but would be inappropriate for a boat trailer.
☐ If the lock will be used indoors or outdoors. If an outdoor lock, will it be protected from the elements?

WARDED LOCKS

Warded locks have limited life spans, particularly when used outdoors. The cheapest locks of this type are only good for a few thousand openings and, when locked, give minimal security.

Most of these locks have three wards, although the cheaper ones have only two. Figure 12-2 illustrates the principle. The key must negotiate the wards before it can disengage the spring bar from the slot in the shackle end. Keys are flat or corrugated. The latter is a mark of the Master Lock Co.

Pass Keys

Figure 12-3 shows how a pass key is cut to defeat the wards. The broad tip of the key opens the spring bar; locks with two spring bars require a key cut as shown in Fig. 12-4.

Some corrugated keys can be reversed to fit other locks by filing it on the back of the blades. Alternately,

Fig. 12-1. Padlock nomenclature.

you can file all the unnecessary metal off, converting the key into a pass key.

In many localities pass keys are illegal and, unless you are a locksmithing student from an accredited school, a locksmith trainee, or a licensed locksmith, possession of such a key is a criminal offense. If you have the need for a pass key, keep it in a safe place in your home or office.

Key Cutting

A warded padlock key is simple to duplicate. Follow this procedure:

Fig. 12-2. A warded padlock. Section A represents a rotating-disc keyway; section B one of the three wards; section C the spring bar.

Fig. 12-3. Wards and their limitations. View A shows the ward arrangement and the necessary key bitting; view B illustrates a pass key.

1. Using the original key as a guide, select the appropriate key blank.

2. Smoke the original key and mount the original and the blank in your vise.

3. Using a 4″ warding file, cut away the excess metal until the blank is an exact copy of the original.

4. Turn the keys over and repeat the operation.

5. Remove the burrs from the duplicate and test it in the lock. Should it stick, the cuts are not deep or wide enough. Make the appropriate alterations and you will have a perfect duplicate key.

Key Impressioning

Impressioning a warded key should take less than five minutes. Follow this procedure:

1. Select the appropriate blank and thoroughly smoke it.

Fig. 12-4. A pass key for a lock with two spring bars.

2. Insert the key and twist it against the wards. Do this several times to get a clear impression.

3. Mount the key in the vise.

4. Make shallow cuts where indicated.

5. Smoke the key again and try it in the lock.

6. Remove the key and file the cuts as indicated.

7. Continue to smoke, test, and file until the key turns without protest. Do not go overboard with the file. If the bits are too deep, the key will work but may break off in the lock.

Repairs to the lock itself are out of the question, since it would be cheaper to buy a new one.

American (Junkunc Brothers)

All American (Junkunc Brothers) padlocks share the same patented locking device—two hardened steel balls fitting into grooves in the shackles. This arrangement is the best ever devised. Applying force on the shackle wedges the balls tighter. The H10 model, for example, requires more than 5000 lb. to force; test locks have been stressed to 6000 lb. and still worked. All these locks use a ten-blade tumbler. The exceptions to these statements are those locks with a deadlock feature. The key must be turned to lock the shackle. These padlocks are made on an entirely different principle and are not discussed here.

American padlocks have a removable cylinder that simplifies servicing. If a customer wants keyed-alike locks, the modification takes only a few minutes. One may also key alike different models.

These locks have three basic subassemblies: cylinder assembly, locking mechanism assembly, and the shackle assembly.

Cylinder Removal and Installation. Follow this procedure:

1. Open the lock, exposing the retaining screw at the base of the shackle hole.

2. Remove the retaining screw (I in Fig. 12-5). If the screw is stubborn, use penetrating oil, letting it set a few moments before attempting to turn the retaining screw.

3. Strike the side of the lock with a leather mallet. The purpose is to force the retaining pin (C) into the space vacated by the retaining screw. Referring to the drawing, note how the pin retains the cylinder in the slot shown.

4. Withdraw the key together with the cylinder.

5. Place a new cylinder in the case.

6. Assemble in the reverse order of disassembly.
Note: The H10 series uses a plated silver cap dimensionally identical to the brass caps on the other models.

The Locking Mechanism. The brass retainer is the heart of the assembly (E in Fig. 12-5). Retainers vary in size according to the lock model. A30 and AC20 retainers have a groove milled in the body. L50, K60, and KC40 retainers are plain. The H10 is much larger than the others.

The retainer assembly has three parts—a washer,

A. Key
B. Cylinder
C. Cylinder-Retaining Pin
D. Retainer Washer
E. Retainer Assembly With Coiled Spring
F. Hardened Steel Balls
G. Case
H. Shackle Spring
I. Retainer Screw
J. Shackle or "Staple"

Fig. 12-5. A popular American padlock.

A. Key
B. Cylinder Assembly
C. Retainer Washer
D. Retainer Assembly
E. Hardened Steel Balls

F. Case or Body
G. 6/32 Brass Screw
H. Shackle Spring
I. Shackle or "Staple"

Fig. 12-6. An American pin-tumbler padlock.

coil spring, and retainer body. The washer (D) fits over the retainer. One end is hooked to accept the coil spring. The other end of the spring is moored in a hole on the retainer.

To remove the retainer, follow this procedure:

1. Grasp the retainer body (E) with needle-nosed pliers.
2. Rotate the retainer 45° and pull.
3. Remove the steel balls.

To replace, follow these steps:

1. Grease the balls so they will stay put.
2. Replace the balls and spread them apart.
3. Exert pressure against the shackle to hold the balls.
4. Determine that the washer is correctly aligned with the spring. The free end of the spring has to be in line with the retaining-pin hole in the case.
5. Place the retainer assembly about halfway in the case, with the washer riding in the grooves provided. The end of the spring also rides in the groove.
6. Twist the retainer about a quarter turn to the right and down.

Shackle. The various models have different shackle lengths and diameters. The spring (H in Fig. 12-5) must match. For example, an L-shaped spring is used on the L50 lock. An extra pin is used to guide the long spring on the K60. The shackle is secured in the H10 with a hardened steel pin; other models secure the shackle with the retainer mechanism.

Pin-Tumbler Padlocks. The American five pin-tumbler padlock has been designed so the cylinders may be quickly changed (Fig. 12-6). Late production 100 and 200 series use the same cylinder assembly and keyway, reducing the inventory load.

Assembly and disassembly of the pin-tumbler padlock is only slightly different from that of the other American padlocks. To change the cylinder, follow this procedure:

1. Open the padlock, exposing the retaining screw at the base of the shackle.
2. Remove the retaining screw with a small screwdriver.
3. Pull the cylinder out of the case. *Note*: Leave the lock unlocked, do not depress the shackle.
4. Insert a new cylinder in the case. Replace the screws, bringing the cylinder almost flush with the case.

To assemble the locking mechanism:

1. Assemble the retainer (D of Fig. 12-6) and the retainer washer (C) with the free end of the spring tightly against the left side projection on the washer.
2. Place the shackle spring (H) in the hole on the end of the shackle (I).
3. Insert the shackle spring into the deep well of the lock body.
4. Drop the balls (E) into the case bottom and move them into the pockets with a small screwdriver. There must be room for the retainer assembly.
5. Depress the shackle so the balls cannot slide back into the center of the case.

6. Insert the retainer assembly (C and D). The retainer washer (C) fits into the elongated hole in the case.

7. Make sure the assembly is bottomed in the hole.

8. Insert a retainer-assembly tool into the cylinder hole so the step on the tool engages the retainer step. Turn the tool clockwise until the lock opens.

9. Install the cylinder assembly, holding it flush with the case bottom. Insert the 6×32 brass screw (G) into the shackle hole and tighten it down snugly.

The American Lock Company will provide the retainer-assembly tool free for the asking.

Because of the close tolerances of American padlocks, they will not operate properly in extreme cold weather unless Kerns ML3849 lubricant is used. Graphite is acceptable in milder climates.

To remove the cylinder in the 600 series pin-tumbler padlocks, follow this procedure:

1. Unlock the padlock and turn the shackle as shown in Fig. 12-7.

2. Depress the spring-loaded plunger (A) with a small screwdriver until the plunger is flush with the wall of the shackle hole. At the same time, pull on the key; withdraw the cylinder.

Master

The Master Lock Company makes a variety of lock types, ranging from simple warded locks to sophisticated pin-tumbler types.

An improved version of the familiar Master padlock has recently been introduced—the Master Super Security Padlock (Fig. 12-8). It was designed to give greater security than standard locks and is recommended for warehouses, storage depots, and industrial plants, as well as around the home and yard. Figure 12-9 is a cutaway view of the mechanism. Salient features include:

☐ A patented dual-lever system to secure the shackle legs. Each lever works independently of the other and are made of hardened steel.

☐ The long shackle leg is tapered to align the locking levers.

Fig. 12-7. The American 600 series padlock.

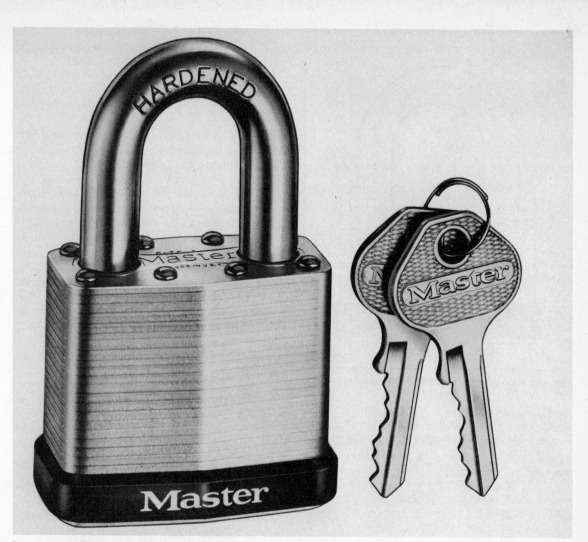

Fig. 12-8. The Master Super Security Padlock.

☐ The case laminations are made of hardened steel and are chrome-plated for weather protection.

☐ The case is larger than that of standard locks for better protection.

☐ A rubberoid bumper prevents the lock body from scratching adjacent surfaces. This feature is not found in other locks.

Tension tests show that this high security lock can tolerate a force of 6000 lbs. on the shackle without damage.

Servicing. Follow this procedure:
1. Drill out the bottom rivets.

2. Remove the plates, one at a time and keep them in order.

3. Remove the lock plug.

4. Assemble the lock using new rivets.

ILCO

ILCO secures the cylinder with what the trade calls the "loose rivet" method. A single brass rivet extends through the case and into the cylinder. The rivet is headless (hence the term "loose") and is secure by virtue of its near invisibility (Fig. 12-10).

To remove the pin, follow these steps:

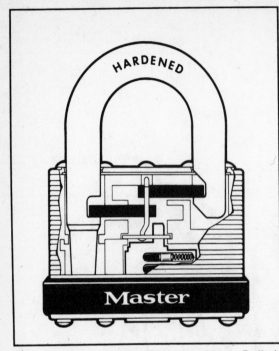

Fig. 12-9. A cutaway of the Master Super Security Padlock.

1. Hold the case up to the light and look for the shadow created by the pin. The pin is located ⅛" from the bottom of the case and is centered on the side.

2. Using a No. 48 bit, drill a hole in the end of the pin.

3. Thread a small screw into the hole.

4. Gently pry up on the screw and remove it together with the pin.

To insert a new pin:

1. Select a piece of brass for stock that exactly matches the diameter of the pin hole.

2. Use the next size smaller rod (to make removal easier if it is ever necessary) and cut it so that 1/32" stands out.

3. Peen over the end of the rod.

4. Burnish the case to camouflage the pin.

Another ILCO variant is a cylinder retaining plate on the bottom of the case. The rivet that holds the assembly together is cunningly disguised in the maker's name stamped on the plate.

To remove the retaining plate, follow this procedure:

1. Draw a line under the ILCO name.

2. At a right angle to the first line, draw a second line (on the outside edge of the O in ILCO).

3. Drill a shallow hole to accept the tip of a 6×32 machine screw.

4. Fit the screw with a washer and thread it into the hole.

5. Using the screwhead for purchase, pry the retainer out of the case.

6. Withdraw the cylinder.

When the cylinder has been serviced and replaced into the case, mount the retainer in its original position, tapping it home with a flat-ended punch. Fill the hole with brass stock.

Kaba Security Padlocks

The Kaba KP 2008 security padlock (Fig. 12-11) is a good quality lock—heavy-duty, with a stainless steel body for a wide variety of applications in the home and industry. It is fitted with an interior dust cover which, when coupled with the stainless steel body, means it can be used in any weather conditions. The dimensions are in Fig. 12-12.

The KP 9 cable padlock has a plastic-coated steel cable. It is very well suited for securing bulky items such as equipment on building sites, bicycles, etc. (Fig. 12-13).

The KP 10 padlock (Fig. 12-14) is designed for strength, the shackle load being over two tons. Dimensions are shown at Fig. 12-15.

The "Epilok" security hasp and staple (Fig. 12-16) is a unique unit. A strong hasp and staple with hardened steel construction and utilizing the unique "Epilok" spigot lock with its advanced Mini-Kaba locking cylinder, the unit will stand up to the most strenuous attack.

Figure 12-17 breaks the unit down a little more. You actually have the hasp unit and the locking spigot lock,

LOOSE-RIVET PADLOCK

Fig. 12-10. ILCO loose-rivet padlock.

Fig. 12-11. Heavy-duty padlock with the Kaba 20 cylinder. (courtesy Kaba Locks, Ltd.)

These last two locks are excellent counter displays for the locksmith that practically sell themselves to the customers. There is no need to provide any form of sales pitch or "hype" for the product. They are top-of-the-line, with proven quality, reliability, and durability, and will give your customer years of valuable protective service.

WAFER-DISC PADLOCKS

Wafer-disc padlocks are recognized by their double-bitted keys and can be a headache to service. If you do not have a key, the lock can be opened by either of two methods:

☐ Picking is possible, but it takes practice. Purchase a set of lockpicks for these locks and spend a few hours learning the skill.
☐ You can try your luck with a set of test keys, available from locksmith supply houses.

Disassembly

If the lock is already opened or a key is available, your job is almost half done.

1. Release the retaining ring clip with a length of stiff wire inserted into the toe shackle hole.
2. Remove the cylinder plug and lay out the parts.

Keys

Duplicate keys can be cut by hand or with a machine

and separate is the rawbolt staple (on the right). Once the hole is drilled and the rawbolt inserted, the staple is screwed on tight, making removal exceptionally difficult. The security provided in this manner is much greater than that of several screws that one would expect with a security hasp. The dimensions are provided at Fig. 12-18.

The 45201 security window/patio door lock has a universal locking mechanism that can be used for any application where a sliding door or window requires additional locking on the top and/or bottom part of the frame. It is also very suitable for hinged windows or pivoting windows (Fig. 12-19).

The lock is operated by a Kaba-20 cylinder mechanism. The same key can be linked to an unlimited number of locks with the same key code, including rim locks, padlocks, or mortise locks. All windows, patio doors, front and back doors can be quickly locked, every time. There is no need to search for individual keys or key on a large key ring. The dimensions for the window/patio door lock are in Fig. 12-20.

Fig. 12-12. Specification data for the heavy-duty padlock. Notice the two different shackle sizes available. (courtesy Kaba Locks, Ltd.)

Fig. 12-13. Cable padlock suitable for securing small pieces of equipment and bicycles. (courtesy Kaba Locks, Ltd.)

Fig. 12-14. The KP 10 has a large shackle into a solid locking body for increased strength. (courtesy Kaba Locks, Ltd.)

designed for this purpose. These machines are expensive. Locksmiths get around the problem by stocking cylinder inserts and precut keys. The insert replaces the original wafer mechanism. A precut key requires that the discs be realigned to fit. You will need some blank discs and should have access to the cutting tool described in the chapter on double-bitted locks and keys.

In extreme cases you can insert the key and file the disc ends for shell clearance. This is not the sort of thing

that a real professional would do, but it works.

Do not attempt to cut double-bitted keys by the impressioning method. A new plug and key is the better choice.

PIN-TUMBLER PADLOCKS

There are hundreds of pin-tumbler locks on the market, but most fall into two categories based on the case construction—laminated or extruded.

Fig. 12-15. KP 10 specification data. (courtesy Kaba Locks, Ltd.)

Laminated Padlocks

Slaymaker pin-tumbler padlocks are made up of a series of steel plates held together by four rivets at the casing corners. Follow these service procedures:

1. Use a hollow mill drill to shear off the rivet heads on the case bottom; this technique allows the rivets to be reused.

2. Remove the bottom plate.

3. Remove the entire cylinder section in one piece.

4. Two cylinder-housing types are used. The most popular requires a follower tool to keep the pins intact.

5. Make the necessary repairs.

6. Insert the cylinder into the casing and replace the bottom plate over the four rivet ends.

7. After checking the action, use a ball-peen hammer and repeen the rivet ends.

Extruded Padlocks

Extruded locks are made from a single piece of metal—usually brass. You will need these tools to service these locks:

☐ A small nail with the point cut off.
☐ Key blanks.

Fig. 12-16. The Epilock high security hasp and staple unit uses the Mini-Kaba cylinder. (courtesy Kaba Locks, Ltd.)

Fig. 12-17. Epilok unit broken down into separate sections for installation. (courtesy Kaba Locks, Ltd.)

☐ Pliers.
☐ A small punch.
☐ Hammer.
☐ A small light (overhead lights are too bright; a flashlight is acceptable).

To disassemble the extruded lock, first locate the plug pins or cover on the bottom of the case. Since the case is highly polished, this is sometimes hard to do. Tilting the lock under the light will show a faint outline of the cover cap or plug pins.

To assemble extruded locks held together by exposed pins, follow this procedure:

1. Determine the diameter of the plug pins.
2. Select a drill bit smaller than the diameter of the pin.
3. Drill slowly and only a fraction of an inch deep—the pin does not extend into the lock very far. If you drill through it you can damage the spring.

4. Extract the plug.
5. Remove each spring and pin.
6. Remove the cylinder. The cylinder is held by a pin that may be covered by a small pin plug. Rapping the lock against a hard surface may shock the plug loose. If not, dislodge the pin plug with a small screwdriver.
7. Make the necessary repairs.
8. Assemble the lock.
9. Fill the pin holes with brass wire.
10. File the wire flush with the case and polish.

Extruded locks with a retaining plate over the pins require a slightly different procedure. If the lock is open, drive the plate out with a punch inserted at the shackle hole. If locked, follow this procedure:

1. Drill a small hole off-center in the plate.
2. Using an icepick or other sharp instrument, drive the plate down, toward the shackle. This will buckle the plate and cause the edges to rise.

Fig. 12-18. Specification data for the Epilock. (courtesy Kaba Locks, Ltd.)

3. Pry the plate loose by working a sharp instrument around the edges.

4. If the plate is not severly damaged, it can be reused. Bow the plate slightly so that the center bulges outward when the plate is installed.

5. Peen the edges of the plate to form a tight seam.

Other locks mount the plug and cylinder by means of horizontal pins running across the width of the lock. These pins can be seen under strong light.

Key Fitting for Pin-Tumbler Padlocks

Follow this procedure:

1. Insert the blank into the plug.
2. Place No. 1 pin into its chamber.
3. Using a blunted nail as a punch, drive the pin into the blank. A single light hammer tap is enough to impression the blank.
4. Remove the blank.
5. File the blank at the impression.

Fig. 12-19. Window or patio door security is assured with this lock unit. (courtesy Kaba Locks, Ltd.)

160

Open position

Closed position

Fig. 12-20. Specification data for the window/patio lock. (courtesy Kaba Locks, Ltd.)

6. Insert the key and try to turn the cylinder. File as necessary.

7. Move to the next pin and repeat the process.

HELPFUL HINTS

It's easier to work with padlocks that are open. If the lock is not open, pick it open. Sometimes the shackle bolt can be disengaged with the help of a hatpin inserted through the keyway. However, locks are getting better and this technique doesn't work as well as it used to.

☐ Polish the lock before returning it to the customer. Use emery paper to remove the deep scratches, then burnish with a wire wheel. Finish by buffing.

☐ The best security in a padlock comes when the shackle is locked at both the heel and toe. The double bolt action (or balls) is the ultimate in padlock security, making it nearly impossible to force the shackle.

☐ When picking fails (and even the best of locksmiths may occasionally have this problem), use penetrating oil. Yale, Corbin, and ILCO pin-tumbler locks are especially susceptible to penetrating oil.

Chapter 13

Home and
Business Services

Over the years a locksmith acquires a great deal of knowledge; much of this information is learned from correcting his own mistakes. This is a hard school. The best way to solve problems is to never let them arise in the first place.

This chapter includes some pretty basic techniques. Some of this material may seem obvious, and you may wonder why it's in here. The answer is that these techniques have been tested and approved by the experts. They will save you time, money, and the embarrassment of callbacks.

The following few pages are supplied courtesy of Corbin (Emhart Corporation). The first section applies to all locksets, regardless of make. Other sections detail service and troubleshooting procedures for specific Corbin locksets, as well as other makes of locks.

COMMON PROBLEMS AND TROUBLESHOOTING

Should you encounter difficulties in the operation of a lockset, I recommend that you first review this checklist of common problems and solutions to see if you can clear up the difficulty.

☐ Is the door locked? (Fig. 13-1)

☐ Are you using the right key? (Fig. 13-2)
☐ Latchbolt or deadbolt does not engage or disengage the strike, or binds in the strike—usually due to bolt-strike misalignment.
☐ Has the door warped? (Fig. 13-3)
☐ Is the door binding? Frames which are out of plumb are frequently the cause of faulty operation of locksets and binding of bolts in the strike (Fig. 13-4).
☐ Are the hinges loose? Tighten the screws, filling holes if necessary, or rehang the door if the screws will not hold (Fig. 13-5).
☐ Are the hinges worn? If excessive wear has occurred on the hinge knuckles, the door will not be held tightly. Replace the hinges.
☐ Is the frame sagging? If sag cannot be corrected and the door and frame returned to plumb relationship, planing or shaving of the door and repositioning or shimming the strike may relieve this condition.
☐ Key operates the latchbolt or deadbolt with difficulty. Usually due to bolt-strike misalignment (Fig. 13-6).

Corbin Cylindrical Locksets

Problem 1—Latchbolt will not deadlock. Caused by deadlocking latch going into strike. Either the strike is

Fig. 13-1. Is the door locked?

Fig. 13-4. Is the door binding?

Fig. 13-2. Are you using the right key?

Fig. 13-5. Are the hinges loose?

Fig. 13-3. Has the door warped?

Fig. 13-6. File to correct bolt-strike misalignment.

out of line or the gap between door and jamb is too great. Realign the strike or shim the strike out towards the flat area of the latchbolt (Fig. 13-7).

Problem 2—Latchbolt cannot be retracted or extended properly. Caused by latchbolt tail and latchbolt retractor not being properly positioned (Fig. 13-8). Remove the lockset from the door. Reinsert the latchbolt in the door. Looking through the hole in the door, the tail should be centered between the top and bottom of the hole. Remove the latchbolt and insert the lockcase. Looking through the latchbolt hole in the lock face of the door, the latchbolt retractor should be centered in the hole. Adjust the outside rose for proper position. Rebore the holes, if necessary, to line up the retractor and tail.

Problem 3—Latchbolt will not project from the lock face (Fig. 18-3). Latchbolt tail and retractor may be misaligned. See *Problem 2*. If this is not the cause, the spring is probably broken.

Fig. 13-7. Latchbolt will not deadlock.

Fig. 13-9. Latchbolt will not project from lock face.

Problem 4—Key works with difficulty. Lubricate the keyway (Fig. 13-10). Do *not* use petroleum products. Spray powdered graphite into the cylinder or place powdered graphite or lead pencil shavings on the key. Move the key slowly back and forth in the keyway. Bitting (notches) on the key may be worn.

Corbin Heavy-Duty Cylindrical Locksets: Service Procedures

Servicing these locks is not difficult if the task is approached methodically.

To Tighten the Locksets:

1. Tighten the inside rose thimble with a wrench. If the thimble needs to be taken up a great deal, tighten the *outside* rose at the same time to prevent possible misalignment and binding of both (Fig. 13-11).

2. If the lockset is still not tight, back off the thimble; using a screwdriver push down the knob retainer and remove the knob and rose. If the spurs on the back of the rose are bent, straighten and reposition the rose so the spurs are embedded in the door. If the spurs are broken

off, insert a rubber band or nonmetallic washer under the rose. Tighten the thimble.

To Remove and Install the Locksets:

1. Remove the key from the knob.

2. Loosen the inside rose thimble with a thimble wrench. Pull the lock slightly to release the rose spurs from the door.

3. Disengage the inside knob retainer with a thimble wrench and pull out the inside knob.

4. Slide the rest of the lockset from the outside of the door.

5. Remove the bolt faceplate screws and slide the latch unit from the lockface of door.

6. To reinstall, reverse procedures above. When inserting case and keyed knob from the outside of the door, be sure the bolt retractor in its case properly engages the latchbolt tail.

To Remove and Replace the Cylinder in the Locksets:

1. Follow the procedure in the preceding section.

Fig. 13-8. Latchbolt cannot be retracted or extended properly.

Fig. 13-10. Lubricating keyway.

Fig. 13-11. Tightening the lockset.

2. Remove one case screw and slightly loosen the other.

3. Swing out the outside knob retainer (it pivots on the case screw slightly loosened above).

4. Using a screwdriver, pry the knob filler cover off the outside knob.

5. Using special Waldes ring pliers No. 3, remove the large Waldes ring from the groove and withdraw the shank and cylinder (Fig. 13-12).

6. To reinstall the cylinder, reverse the above procedure. Be sure the knurled side of the Waldes ring is face up.

To Change the Hand of the Locksets:

Remember that the lock cylinder should always be in position to receive the key with the bitting (notches) facing upward.

1. Follow the procedure in the preceding section.

2. Slightly loosen *one* case screw and back off the other (Fig. 13-13).

3. Swing out the outside knob retainer (it pivots on case screw slightly loosened above).

Fig. 13-13. Changing lockset hand.

4. Lift the knob out and rotate it 180°. Replace it in case. Swing the knob retainer back into place.

5. Insert the case screw and tighten both screws.

6. Reinstall (Fig. 13-14).

Corbin Standard-Duty Cylindrical Locksets: Service Procedures

To Tighten the Locksets:

1. Depress the knob retainer and remove the inside knob.

2. Unsnap the rose from the rose liner (Fig. 13-15).

3. Place the wrench into the slot in the rose liner and rotate clockwise until tight on the door. If the lockset is extremely loose, tighten the outside rose and inside liner *equally* (Fig. 13-16).

4. Replace the inside rose; depress the knob retainer and slide the knob onto the spindle until the retainer engages the hole in the knob shank.

To Remove Locksets:

To remove, follow this procedure:

1. Depress the knob retainer and remove the inside knob.

Fig. 13-12. Removing cylinder.

Fig. 13-14. Reinstalling knob.

Fig. 13-15. Unsnapping rose from rose liner.

2. Unsnap the rose from the rose liner.

3. Place the wrench into the slot in the rose liner and rotate counterclockwise until disengaged from the spindle.

4. Remove the lock by pulling the outside knob.

5. Remove the bolt after removing the screws in the lockface of the door.

To Reinstall Locksets:

1. Adjust the lock for the door thickness by turning the outside rose until the edge of the rose matches one of the lines marked on the shank of the knob. (First line is for 1⅜" door, second for 1¾" door). When using trim rosettes, increase the adjustment to compensate for the thickness of the metal.

2. Install the latchbolt in the face of the door.

3. Install the lock from the outside of the door so the case engages the slot in the latchbolt and the tail interlocks the retractor.

4. Replace the rose liner, rose, and knob on the inside of the door.

Fig. 13-17. Locking exterior knob.

To Reverse the Knobs:

1. Lock the exterior knob (Fig. 13-17).

2. Hold the latchbolt in the retracted position with the key (Fig. 13-18). Depress the knob retainer with a screwdriver through the slot in the knob shank.

3. Pull the knob off the spindle (Fig. 13-19). Rotate the knob so the bitting on the key will be *up*.

4. Replace the knob by lining up the lance in neck of knob with the slot in the spindle.

5. Push the knob on the spindle until it hits the retainer button. Depress the retainer button, and push the knob until it snaps into position.

To Remove the Cylinders:

1. Remove the outside knob. Turn the key in either direction until it can be partially extended from the plug (Fig. 13-20).

2. Hold the knob and turn the key to the left, pulling slightly on the key until the cylinder disengages.

Fig. 13-16. Tightening rose.

Fig. 13-18. Retract latchbolt.

Fig. 13-19. Pull knob off spindle.

Troubleshooting Corbin Unit Locksets (300 and 900 Series)

Problem 1—Latchbolt binds or rattles in strike. Adjust the strike. Because the lockset is preassembled as one unit, there are no internal adjustments to be made. Check to be sure that the cutout for the lockset is square and at the right depth so the face of the lockset is flush with the face of the door. Once this has been done, adjust the nylon adjusting screw in the strike (Fig. 13-21).

Problem 2—Key does not activate the knob or latchbolt. Check for worn key. If bitting (notches) in the key is worn down or the key is bent, the locking mechanism will not operate properly. If the key is not worn, spray powdered graphite into the keyway or put graphite or lead pencil filings on the key and move it back and forth slowly in the keyway (Fig. 13-22). Never use petroleum products. Check to be sure that a binding latch is not the cause.

Problem 3—Lockset is loose in the door. Tighten the escutcheon screws (Fig. 13-23). Be sure to tighten evenly.

Problem 4—Lockset has the wrong bevel for the door. Reverse the lockset. Unit locksets with horizontal keyways may be changed from right to left hand regular bevel, or vice-versa—or from right hand reverse bevel to left hand reverse bevel, or vice-versa—by merely turning the lock upside down (Fig. 13-24).

Fig. 13-21. Adjusting screw in strike.

Corbin Unit Locksets (300 and 900 Series): Service Procedures

Usually the unit locksets are easier to service than the cylindrical locksets. To service this series of locksets proceed as follows:

To Remove the Locksets From a Door:

1. Remove the through-bolts on the inside of the door (Fig. 13-25).

2. Push the outside escutcheon away from the door so the lugs clear the holes.

3. Slide the assembly out of the door.

Fig. 13-20. Remove outside knob.

Fig. 13-22. Lubricating keyway.

Fig. 13-23. Tightening escutcheon screws.

To Remove 900 Series Knobs:

1. Remove the attaching screws (Fig. 13-26).
2. Snap off the dust cover.
3. Pry the wire retaining ring from the knob retaining key located in a slot in the frame tube. Remove the retaining key. Remove the inside knob which is fastened to the knob shank.
4. Remove the inside escutcheon.
5. Loosen the escutcheon on the outside of the lock by inserting a screwdriver through the access hole from the inside of the lock frame and remove the screw and escutcheon fastener.
6. Pry the wire retaining ring from the knob retaining key located in a slot in the frame tube. Remove the retaining key. Remove the outside knob which is fastened to the knob shank.

To Remove Cylinders:

1. Using a screwdriver, pry the knob filler cover off the outside knob.

2. Use No. 103F92 Waldes retaining pliers to remove the Waldes ring.
3. Remove the shank and lock cylinder. Rekey the lock.
4. Reverse the procedure to reinstall. Be sure that the beveled edge of Waldes ring faces away from the knob and that the ring is properly seated in the groove (Fig. 13-27).

To Remove 300 Series Knobs:

1. Remove the key from the lock (lock should be unlocked). Remove the three attaching screws (Fig. 13-26).
2. Snap off the dust cover.
3. On the inside knob side, pry the wire retaining ring away from the knob retaining key located in the slot in the frame tube. Remove the retaining key. Remove the knob that has the shank assembled to the knob.
4. Remove the inside escutcheon.
5. Loosen the escutcheon on the outside of the lock by inserting a screwdriver through the access hole from the inside of the lock chassis and remove the screw and washer.
6. Repeat step 3 on the outside knob and shank.

Troubleshooting Corbin
Mortise Locksets (7000, 7500, 8500 Series)

Problem 1—With door open, latchbolt doesn't extend or retract freely. Check for binding against the rose. Adjust the knob. Loosen the roses or trim on the door. If the bolt now operates freely, the roses or trim must be realigned. A knob aligning tool is recommended. Check installation templates for proper position (Fig. 13-28). If the bolt does not operate properly with trim and roses loosened, remove the lockset from the door. If the lockset operates

Fig. 13-24. Reversing lockset.

Fig. 13-25. Removing through-bolts.

Fig. 13-26. Removing attaching screws.

properly when removed from the door, use a chisel to make the mortise larger so that the lockset enters freely.

Problem 2—With door closed, latchbolt doesn't extend or retract freely, or door won't latch at all. Open the door. If the latchbolt still doesn't operate properly, see *Problem 1.*

Problem 3—Latchbolt "stubs" on the strike lip. Bend the strike slightly back toward the jamb. Wax or paraffin makes an excellent lubricant, as does silicone spray (Fig. 13-29).

Problem 4—Deadbolt doesn't enter the strike. This is probably due to misalignment of strike and bolt, particu-

larly in cases where door sag has taken place. Both latchbolt and deadbolt holes in the strike must be filed or the strike repositioned. Do *not* force the thumbpiece if the deadbolt doesn't extend and retract in the strike freely (Fig. 13-30).

Fig. 13-27. Removing cylinder.

Fig. 13-28. Checking latchbolt.

169

Fig. 13-29. Bending strike.

Fig. 13-31. Removing cylinder.

Problem 5—To remove the cylinder. Loosen the cylinder locking screws in the face of the lockset. If the scalp covers the set screws, remove the scalp. Unscrew the cylinder. When replacing, be sure the locking screws are firmly seated (Fig. 13-31).

Problem 6—Key does not operate the latchbolt or deadbolt. Loosen the cylinder locking screws. The cylinder is in the wrong position in the door so the cylinder cam does not engage the locking mechanism properly. Turn the cylinder a whole turn to the left or right until it works properly. The keyway must always be in position to receive the key with bitting (notches) up. Tighten the cylinder locking screws.

Problem 7—Key turns hard when retracting the deadbolt. If the bolt operates freely with the door open, check the bolt-strike alignment. Check the scalp to be sure it is not binding the bolt, or that paint over the bolt is not causing the bind.

Problem 8—Key works hard in the cylinder. Lubricate the cylinder with powdered graphite or place graphite on the key and move it back and forth slowly in the keyway. Never use any petroleum lubricants (Fig. 13-32).

Problem 9—Key breaks in the lock. Remove the cylinder from the lock. Insert a long pin or wire into the back end (cam end of the cylinder). Move it back and forth until the broken key stub is forced out through the front of the cylinder. Clean the cylinder with ethyl acetate and lubricate it with graphite before reinstalling (Fig. 13-33).

Problem 10—Thumbpiece trim doesn't retract latchbolt completely or doesn't extend bolt completely. Check for binding at inside trim; or if the outside thumbpiece trim is used in conjunction with the inside panic device, check to see if it is operating properly and is not dogged down. Remove the thumbpiece and check the position of thumbpiece in relation to the latch trip at the bottom of the mortise lock case. When properly installed, the top of the thumbpiece should be up against the bottom of the latch trip, but not lifting it (Fig. 13-34). If the bolt doesn't retract fully when the thumbpiece is pushed down, the thumbpiece is too low on the door. Move the trim up as needed. If the bolt doesn't extend completely when the

Fig. 13-30. Filing strike.

Fig. 13-32. Lubricating keyway.

Fig. 13-33. Lubricating cylinder.

thumbpiece is released, the thumbpiece is too high on the door. Move the trim down as needed. If the trim is fixed and cannot be moved, carefully bend the thumbpiece tail up or down.

Corbin Mortise Locksets (7000, 7500, 8500 Series): Service Procedures

Mortise locksets are not difficult to work with. Servicing them involves only a few basic procedures.

Fig. 13-34. Installing thumbpiece.

Fig. 13-35. Removing spindle from knob.

To Install and Adjust the Working Trim:

1. Unscrew the spindle. Remove the mounting plates, roses, and thimble from the spindle. Unscrew the sleeve from the inside knob. Remove the spindle from the knob (Fig. 13-35).

2. Install the rose attaching plates. On wood doors install flange washers through the mortise, turning the attaching plates into them. It may be necessary to mortise for the washer to permit the mortise lock to clear it. Install the mortise lock.

3. Align the plates using a No. 028 aligning tool. If not available, assemble the spindles allowing 1/16″ gap between halves of the swivel spindles, with mounting plates, sleeve spacer, and adjusting nut in position shown in illustration. Tighten the adjusting nut with fingers. Mark screw holes through both mounting plates. Install the screws.

4. Reassemble the spindle, tightening the adjusting nut, and back off ¼ turn to line up with the flats of the spacer. The spindle should have a slight end chuck. Try the knob. If the latch binds, back off the adjusting nut another ¼ turn. Disassemble the spindle.

5. Assemble the roses and thimble. Reassemble the spindle complete with knobs. Tighten the rose thimbles and inside adjusting plate with furnished spanner wrenches. Knobs should turn freely from either side in either direction.

**To Install and Adjust
the Lever Handle Trim Assembly:**

1. Check to be sure that the handle hole is 1⅛″ wide. On wood doors, remove the mortise lockset and make additional mortise for retainer on the inside of the case mortise. Replace the locksets and secure with screws.

2. Place the rose assembly spindle into the lock hub. The label shows top and latch edge. Spot four screw holes through the assembly plate (Fig. 13-36). Make sure that the latchbolt operates freely. Drill holes very carefully.

3. Attach the rose assembly to the door. (With a wood door, attach through door to retainer.)

4. Put on the cover plates. Place the lever handle on the assembly and secure with an Allen set screw.

5. Adjust. If the latchbolt binds, remove the lower plate and slightly change rose assembly position.

Fig. 13-36. Installing lever handle.

Fig. 13-37. Reversing the hand.

To Reverse the Hand:

As you will remember from Chapter 11, the *hand* refers to the position of the hinge on the door and to the direction of swing. Some locksets are universal and fit all four hands. Others must be modified in the field.

1. Refer to Fig. 13-37. Unfasten the two cap screws (A) and remove the cap.

2. Reverse the latchbolt (B). If the lock has an antifriction latch, first remove the L-shaped pin (C) and reinstall it on the opposite side after reversing the bolt. The short leg should be held against the side of the case with the pressure-sensitive decal.

3. Reverse the auxiliary latch (D) if an auxiliary latch is furnished. To do this first remove the auxiliary latch lever and spring, then invert the latch so the concave surface is toward the stop in the frame, then reassemble the lever and spring.

4. Reverse the front bevel. To do this loosen three

screws (E) in the top and bottom edges of the case, adjust to the proper bevel (or to flat), and retighten the screws securely.

5. Reverse the knob hubs (F): hub with ⅜″ square spindle hole toward the outside.

6. Replace the cap and securely tighten the screws.

7. Check for proper operation before installing in the door.

Mounting the Medeco Ultra 700 Deadbolt

Instructions for mounting the Medeco Ultra 700 deadbolt are as follows:

1. Cut the door using the template as a guide (Fig. 13-38).

2. Place the wave spring inside the cylinder guard.

3. Insert the cylinder through the cylinder guard with the wave spring *under* the head of the cylinder.

4. Thread the cylinder into the lock body securely. The keyway must be horizontal and on the side of the cylinder closest to the bolt. If the keyway does not come to the horizontal, loosen the cylinder as necessary (always less then one turn).

5. Remove the faceplate and install the four set screws at the top and bottom of the lock.

6. Secure the cylinder with the set screws provided.

7. If you have to disassemble the lock, insert a wrench into the hole and turn counterclockwise. This will open the lock.

8. With the lock near or full open, manipulate the lock so the pins slip out of one escutcheon.

9. Reverse this procedure to reassemble. *Note*: The 3/16″ roll pin must fit into the hole in the opposite escutcheon for alignment.

10. With the lock assembled and the cover in place, slide the unit into the door cutout (Fig. 13-39). Position it

Fig. 13-38. Cut door with template.

Fig. 13-39. Slide lock into cutout.

so that the longest escutcheon is flush with the high edge of the door bevel.

11. Tighten the lock to the door by inserting an Allen wrench in the appropriate holes and turning clockwise. This brings the escutcheons of the lock together so they clamp the door.

12. Turn the mounting screws finger tight, then give them approximately ½ turn more (Fig. 13-40).

13. Install the faceplate with the screws provided (Fig. 13-41).

The Ultra 700 lock can be adapted for a single cylinder on the exterior side and a thumbturn on the interior side of the door. In order for the thumbturn to act as an indicator, the long portion (the actual turning mechanism) must point in the direction of the bolt (extended or retracted position) and the cam must, therefore, be properly positioned on the thumbturn before the entire assembly is installed in the lock.

Check the cam position. If necessary remove the drive pin and cam, rotate one half turn (180 degrees), and replace the pin. Figure 13-42 shows the thumbturn as viewed from the interior door side. When installing the lock, be sure the cam is inserted between the mechanism arms, then tighten the cylinder set screws on the top and bottom of the lock.

The instructions are simple, and easy to follow; from a locksmith's viewpoint, the time is well spent in terms of customer relations and business income. In addition, if you are replacing the interior cylinder with the thumbturn, you may well be able to keep the cylinder that was removed, thus adding various parts to your supply stock. (*Note: Never* resell a used lock; the various individual parts of the lock can be used for repair parts, for study, and also for various practical training exercise for new employees.)

Continuing with the MAG Ultra 700 deadbolt, a new series from MAG, the MR series, is an adaption of the original, and one that is well worth investing in. The adaption is for use with the Corbin Master Ring cylinder and is now available from your locksmithing distributors nationwide.

The adaptation (Fig. 13-43) enables the same master keying system to be utilized while still providing the high security lock compatible with the present system.

Like the regular Ultra 700, this new series includes no attaching screws on either side of the lock and the key cannot be withdrawn from the lock unless the locking bolt is either fully extended or fully retracted. Also, the cylinder guard is free-turning and recessed into the escutcheon casting for maximum protection.

A new Series "A" deadbolt Ultra 700 is now also available for the locksmith to add to potential sales. The Series "A" is designed for use where one side entrance only is desired (Fig. 13-44). The blank side of the lock is a soldering casting. The key side of the lock is available in stainless steel and uses a 1⅛" mortise cylinder.

Let's move on to another MAG lock, the Ultra 800, also a deadbolt. The cylinder used with the lock is a 1 5/32" diameter by 1⅛" long mortise cylinder with a standard Yale cam. The lock is set by the factory for 1¾" doors, but for other door thicknesses, instructions are included, and are also below.

Fig. 13-40. Tightening mounting screws.

Fig. 13-41. Installing faceplate.

Long portion points in direction of bolt (extended or retracted)

Cam

Thumb Turn Housing

Drive Pin

Cam points up for mounting on short escutcheon

Cam points down for mounting on long escutcheon

Fig. 13-42. Cam direction positioning when mounting is dependent upon whether a short or long escutcheon is being used. (courtesy M.A.G. Engineering and Mfg., Inc.)

Figure 13-45 illustrates the lock in an exploded view. Notice the bolt; it is a cylindrical deadbolt instead of the standard rectangular bolt mostly associated with deadbolt locks. Also take note of the strike; again while looking like the average mill-run strike, MAG has gone further and insured that it is ably secured to the frame and supporting stud. Also, from the illustration you can readily see that the interior side of the door will take either a mortise cylinder or thumbturn, providing your customer with an option feature.

First, you must determine the proper setting for the lock if the door is not 1¾″ in thickness. Figure 13-46 details adjusting the exterior lock housing piece for use on doors of varying thicknesses. The ring referred to is the circular ring on the reverse side (inside portion) of the exterior lock housing. By looking at the reverse side, at the 8 o'clock position, you locate the hole for the pin to be inserted. With this done, you now proceed to the door preparation, using the template furnished with the lock unit.

Fig. 13-43. The Ultra 700 deadbolt adaption for use with the Corbin Master Ring Cylinder. (courtesy M.A.G. Engineering and Mfg., Inc.)

Fig. 13-44. Ultra 700 deadbolt Series "A" used when one side entrance only is desired. (courtesy M.A.G. Engineering and Mfg., Inc.)

175

Fig. 13-45. Exploded view of the Ultra 800 deadbolt. (courtesy M.A.G. Engineering and Mfg., Inc.)

Door Preparation (wood doors):

1. Follow the dimensions outlined on the template. Drill ⅛″ pilot hole through the door face, and then drill a 2⅛″ hole exactly 2¾″ from the high edge of the door bevel, referring to the template as you do so.

2. Drill a 1″ hole through the edge of the door, breaking into the 2⅛″ hole. Cut out the wood for the bolt housing and faceplate 1⅛″ wide by 2¼″ high by 3/32″ deep. This faceplate hole must be centered on the 1″ hole.

3. Drill the four ⅛″ pilot hole for the four corner mounting bolts. With the pilot holes drilled, replace the ⅛″ drill bit with a 9/16″ bit and proceed to put the four 9/16″ hole through the door. *Note:* MAG has a drill fixture kit available for fast and accurate installation. If you expect to be performing numerous installations, then the drill fixture kit is a must-have item that becomes a necessity for rapid installation.

Prior to mounting the lock, you should assemble the mortise cylinder to the exterior escutcheon of the lock, as follows. (The numbers refer to the various lock part components in Fig. 13-45.)

1. Put the cylinder spring (2) over the cylinder (1)

then insert the cylinder into the cylinder guard cover (3), and install the escutcheon cover (12).

2. Now thread the cylinder in until the cylinder is tight (the keyway should be toward the rear of the lock and horizontal). Tighten the set screw. The cylinder guard cover should be tight against the escutcheon cover. The ring should rotate freely one-half turn back and forth. If the cam on the cylinder touches inside the mechanism cover, back off the cylinder one turn.

3. Put the key in the cylinder and rotate one complete turn, first in one direction and then the opposite direction. This is to check the function of this assembly.

4. If you haven't adjusted the lock for other than a 1¾″ door, remove the mechanism cover by first removing the spring assembly. Pry up the cover and remove the ring and then adjust the pin for the appropriate door thickness, referring to Fig. 13-46 to ensure the details are correct.

Mounting the Lock (wood doors):

1. Insert the bolt assembly into the 1″ hole.

2. Insert the exterior escutcheon (with the mechanism ring in the retracted position) into the 9/16″

RING

STANDARD 1"
PIN

OPTIONAL 1 1/2" PIN
FROM SCREW PACKAGE
PART NO.862 FOR
1 7/8" TO 2 1/4"
DOORS

A

B

DOOR THICKNESS	A
1 3/8	.500
1 1/2	.600
1 5/8 AND 1 3/4	.800

DOOR THICKNESS	B
1 7/8	1.000
2	1.100
2 1/8	1.250
2 1/4	1.300

ALL DIMENSIONS SHOWN ARE IN INCHES

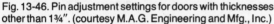

Fig. 13-46. Pin adjustment settings for doors with thicknesses other than 1¾". (courtesy M.A.G. Engineering and Mfg., Inc.)

and 2⅛″ holes. When inserting, make sure the bolt link fits over the mechanism pin and that the *large tab* of the bolt link is towards the hole in the mechanism cover.

3. Place the interior escutcheon on the door and secure it to the exterior escutcheon with the four ¼″ diameter bolts furnished with the lock unit.

4. Put the cover on the interior escutcheon.

Single Cylinder. Install the thumbturn with the *bolt retracted*. Place the circular wave spring over the thumbturn housing. Be sure the projection on the thumbturn cam and the projection on the housing are together. Place the thumbturn assembly into the interior escutcheon. Make sure the alignment projections line up with the bolt in the escutcheon hole. The thumbturn must be pointing away from the edge of the door. Push the assembly in carefully, making certain the hole in the thumbturn cam lines up with the mechanism pin. Push it in until the catch enters the slot in the thumbturn housing. Check for proper function by rotating the thumbturn and by pulling to be sure the catch has engaged the housing. The long side of the thumbturn should point toward the edge of the door when the bolt is extended, and away from the door edge when retracted.

Double Cylinder. Assemble the mortise cylinder, cylinder spring, and cylinder guard cover to the interior cylinder housing the same as the exterior escutcheon. The keyway must be horizontal and opposite the catch slot. Thread the interior ring onto the cylinder until it stops against the cylinder housing, then back off one to one and one-half turns prior to assembly into the lock.

Install the interior cylinder housing assembly with the bolt retracted. Insert the assembly into the escutcheon hole, lining up the boss with the slot. Also line up the ring boss with the escutcheon slot. Press it in fully with the key partially withdrawn, then rotate counterclockwise to stop. Push in on the cylinder until the catch snaps into the slot. Try to turn the cylinder housing clockwise to be sure the catch is engaged to the lock cylinder housing in place. Check for the proper function of the lock, with the key, from both sides.

6. Insert the bolt housing assembly into the 1″ bolt hole. *Caution:* Make certain the opening side of the bolt housing is *up* on left-hand doors and *down* on right-hand doors.

7. Install the faceplate with the screws provided. Make sure the hole in the faceplate fits over the projection of the bolt housing assembly to center it properly.

Metal Doors:

Door preparation and procedures for mounting the lock to hollow metal doors—for doors that have no seam or hole in the center of the door edge—is as follows:

1. Follow the dimensions outlined on the template. Drill ⅛″ pilot hole through the face of the door located 2¾″ away from the high edge of the door bevel. Be sure the pilot hole is the same distance from the high edge of the door on each side. Bore a 2⅛″ hole through.

2. Drill four 9/16″ holes the same as for wood doors.

3. Using the template, make three centerpunch marks on the edge of the door; one in the center, one 13/16″ above and one 13/16″ below center. Drill ¼″ holes.

4. Open the center hole with a ¾″ hole saw. Remove the burr on the inside edge. Countersink top and bottom holes for 12-24 flathead machine screws.

5. Install clip nuts to the bolt housing with the long side toward the rear of the housing.

6. Insert the bolt housing through the 2⅛″ hole in the door and place against inside front edge of the door, making certain the projection of the bolt housing assembly fits into the ¾″ hole. Put opening on side of housing *up* on left hand door and *down* on right hand *doors*.

7. Secure the bolt housing to edge of the door with the two 12/24 screws.

8. Insert the bolt assembly into the bolt housing.

9. Follow steps 2 through 5 for wood doors to complete the installation.

Fig. 13-47. Door edge cutout required for metal door with center seam on the door edge. (courtesy M.A.G. Engineering and Mfg., Inc.)

Fig. 13-49. Properly mark around the strike as closely as possible. (courtesy M.A.G. Engineering and Mfg., Inc.)

For hollow metal doors that have a seam or fold on the edge that would interfere with drilling of holes for mounting the bolt housing:

1. Cut out door edge as shown (Fig. 13-47) and mount Install-A-Lock. (Depending on whether the door is a 1⅜″ or 1¾″ door, you will require a special order; check with your distributor to see what the Install-A-Lock special order number will be for this unit.)

2. Finish the door preparation for hollow metal doors as previously described.

3. Mount the lock as per procedures for mounting to wood doors.

Strike Installation
Wood Jambs:

1. With the Ultra 800 properly installed, close the door fully and throw the bolt out (by either the key or thumbturn) so that the point on the end of the bolt will make a mark on the jamb. This will be the center mark for the strike mounting.

2. Open the door. Place the strike plate against the door, ensuring that it lines up with the bolt mark centered on the strike plate hole (Fig. 13-48).

3. Mark the center hole required to be drilled; also mark around the strike plate edge (Fig. 13-49).

Fig. 13-48. Line up the strike plate to the locking bolt. (courtesy M.A.G. Engineering and Mfg., Inc.)

Fig. 13-50. Ensure the midpoint is centered in the middle of the strike plate hole. (courtesy M.A.G. Engineering and Mfg., Inc.)

Fig. 13-51. Strike box inserted into the door frame. (courtesy M.A.G. Engineering and Mfg., Inc.)

Fig. 13-53. Side view of the strike box. (courtesy M.A.G. Engineering and Mfg., Inc.)

4. Drill two 1″ diameter holes 1⅜″ deep, one ⅝″ above and one ⅝″ below the center mark (Fig. 13-50).

5. Chisel out the wood inbetween.

6. Cut out the wood 1¼″ wide by 4⅞″ high by ⅛″ deep for the strike box and strike plate.

7. Be sure to drill ⅛″ pilot hole for all four screws to aid installation.

8. Insert the strike box. Place the finished strike plate over the strike plate box and secure with screws provided (Fig. 13-51).

9. Within the strike box, drill two ⅛″ diameter pilot holes on approximately a 25° angle toward the center of the jamb. This will provide for maximum holding power when the two screws are attached to the stud (Fig. 13-52). A side view of the strike box positioning is at Fig. 13-53.

Metal Jambs Not Prepared for a Strike:

1. Mark the jamb with the point on the end of the

bolt as described for wood jambs.

2. Drill a ⅞″ hole only. No strike plate is required.

If desired by the customer, an Adjust-A-Strike may be installed with the unit. Several models are available. Figure 13-54 refers to the models 450/451—1¼″×2¾″ lip and deadbolt strike jamb preparation (Fig. 13-55 provides positioning of parts and the strike installation completed for all models described here.)

1. For 1¾″ door, start cutout 3/16″ away from the door stop; for 1⅜″ doors, start the cutout on the edge of the door stop.

2. Cut out 1¼″×2¾″ dimension, 5/32″ deep.

3. Cut out 1½″ dimension, 3/32″ deep.

4. For 1¾″ door, 31/32″; for 1⅜″ door, 25/32″.

5. For latches (lip strike), drill ⅝″ deep; for deadbolts, drill 1″ deep.

6. Install the strike and keepers in the center of the cutout. Close the door to check fit. Adjust if necessary.

Fig. 13-52. Ensure the screws go into the frame stud for maximum holding power. (courtesy M.A.G. Engineering and Mfg., Inc.)

Fig. 13-54. Exploded view illustrating parts relationship for the complete installation of the Adjust-A-Strike. (courtesy M.A.G. Engineering and Mfg., Inc.)

Fig. 13-55. Jamb preparation for the Adjust-A-Strike models 450 and 451. (courtesy M.A.G. Engineering and Mfg., Inc.)

For the model 452, for a 1¼"×4⅞" lip strike, the jamb preparation is as follows:

1. For 1¾" door, start the cutout 3/16" away from the door stop; for 1⅜", door start cutout on the edge of door stop (Fig. 13-56).
2. Cut out 1¼"×4⅞" dimension ⅛" deep.
3. Cut out 3⅜" dimension 1/16" deep.
4. For 1¾" door, 31/32". For 1⅜" door, 25/32".
5. Install the strike and keepers in the center of the cutout. Close the door and check for proper fit, adjusting if necessary.

The Schlage "G" Series Lockset

The mechanics of this lockset have been discussed in an earlier chapter.

Installation:

These instructions, developed from material supplied by Schlage, apply to 1⅜" to 2" wood doors.

1. Mark the height line (the centerline of the deadlatch) on the door (Fig. 13-57). It is suggested that the line be 38 inches from the floor.
2. Mark the centerline of the door thickness on the door edge.
3. Position the template (supplied with the lockset) on the door with the lower deadlatch hole on the height line. Mark center points for one 2⅛" hole and two 1½" holes through the template (Fig. 13-57). Mark the centers for two 1" latch holes on the door edge.
4. Bore one 2⅛" hole and two 1½" holes in the door panel.
5. Bore two 1" holes in the edge of the door. The upper hole should be extended ⅜" beyond the far side of the middle hole on the door panel (Fig. 13-58). Mortise the edge of the door for the latch front.
6. Mark vertical and height lines on the jamb exactly opposite the center point of the lower latch hole. Mark a second horizontal centerline 2" above the height line for the deadbolt hole. Bore a 1" diameter hole 1⅛" deep ¼" below the height line. Bore a second hole to the

Fig. 13-56. Model 452 for 1¾" door, jamb preparation and completed installation. (courtesy M.A.G. Engineering and Mfg., Inc.)

Fig. 13-57. Marking holes.

Fig. 13-58. Boring holes.

Fig. 13-59. Clear area between holes.

Fig. 13-60. Disassemble lock.

Fig. 13-61. Insert latch unit.

Fig. 13-62. Rotate deadlatch.

same dimensions ¼″ above the second horizontal line. Clear out area between holes for the strike box (Fig. 13-59).

7. Install the strike (it can be reversed for either hand). If you have done the work correctly, the strike screws will be on the same vertical centerline as the latch screws.

8. Disassemble the lock. Remove the inside knob and lift off the outside mechanism (Fig. 13-60).

9. With the deadbolt thrown and the crank slot in the vertical position, insert the latch unit and secure with the screws provided (Fig. 13-61). The deadlatch can be rotated to match the door hand (Fig. 13-62). The beveled edge of the deadlatch should contact the striker as the door swings shut.

10. Install the inside mechanism with the knob button released and the crank bar in the vertical position. Insert the end of the crank bar in the slot on the deadbolt (Fig. 13-63). Engage the jaws of the slide with the deadlatch bar, and engage the deadlatch housing with the ears on the inside mechanism. This sounds more confusing than it is; see the insert for Fig. 13-63.

11. From the outside of the door, turn the cylinder bar slot with a screwdriver to check the action of the

Fig. 13-63. Install mechanism.

Fig. 13-65. Rotate cylinder bar.

deadbolt and deadlatch. Once you are satisfied that the lock works properly, retract the deadbolt and release the knob button. Turn the cylinder bar slot to the horizontal position (Fig. 13-64). The clutch plate should be situated as shown.

12. Rotate the cylinder bar to the horizontal position with the knob spindle as shown (Fig. 13-65).

13. Insert the cylinder bar into the bar slot in the top hole. Engage the clutch plate with the outside knob spindle. The spindle must be positioned as shown in Fig. 13-65, and the clutch as shown in Figs. 13-64 and 13-66.

14. G85PD locksets require that the cylinder driver bar be flush with the end of the upper bushing (Fig. 13-67).

15. Install the inside rose with the turn unit in the vertical position and the V-notch on the inside of the turn

pointing to the edge of the door (Fig. 13-68). Engage the top of the rose with the mounting plate and, depressing the catch, snap the rose into place. Align the lug on the knob with the slot in the spindle and depress the knob catch. Push the knob home.

16. Test the lock.

Changing the Hand:

"G" series locksets are available in right- and left-hand versions. However, the hand can be changed in the field.

Figure 13-69 is an assembled view of a right-hand unit. To convert it to left-hand operation, follow these steps:

1. Disengage the retaining rings (Fig. 13-70).

Fig. 13-64. Testing lock.

Fig. 13-66. Positioning clutch.

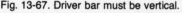

Fig. 13-67. Driver bar must be vertical.

Fig. 13-68. Install inside rose.

2. Lift off the mounting-plate cover (Fig. 13-71).

3. Remove the linkage-bar plate, linkage bar, bolt bar, driver bar, knob-driver linkage arm, and lower bushing (Fig. 13-72).

4. With a factory-supplied wrench or long-nosed pliers, straighten the tabs (Fig. 13-73).

5. Compress the slide with a screwdriver and pull the assembly out (Fig. 13-74).

6. Rotate the spindle 180° to relocate the knob catch. (Fig. 13-75).

7. Rotate the slide assembly 180°. With the slide compressed, insert the lugs into slots on the mounting plate (Fig. 13-76).

8. Bend the tabs to secure the mounting plate (Fig. 13-77).

9. Install the linkage arm (Fig. 13-78).

10. Install the lower bushing—the slot is vertical (Fig. 13-79).

11. Install the retaining ring (Fig. 13-80).

12. Install the driver bar—the bar is horizontal (Fig. 13-81).

13. Install the bolt bar (Fig. 13-82).

14. Install the linkage bar (Fig. 13-83).

15. Install the knob driver (Fig. 13-84).

16. Install the linkage-bar plate (Fig. 13-85).

17. With the upper bushing slot in the horizontal position, install the cylinder driver (Fig. 13-86).

Fig. 13-69. Right-hand unit.

Fig. 13-70. Changing hand, Step 1.

Fig. 13-72. Changing hand, Step 3.

Fig. 13-71. Changing hand, Step 2.

Fig. 13-73. Changing hand, Step 4.

18. Replace the mounting-plate cover (Fig. 13-87).

19. Install the inside retaining rings (Fig. 13-88).

20. Turn the slot in the G85 unit to face the same direction as the slide (Fig. 13-89).

21. Turn the deadlatch 180° (Fig. 13-90).

22. Check your work (Fig. 13-91).

Figure 13-92 is an assembled view of a left-hand unit. To convert it to right-hand operation, follow these steps:

1. Disengage the retaining rings (Fig. 13-93).

2. Lift off the mounting-plate cover (Fig. 13-94).

3. Remove the linkage-bar plate, linkage bar, bolt bar, driver bar, knob-driver linkage arm, and lower bushing (Fig. 13-95).

Fig. 13-74. Changing hand, Step 5.

Fig. 13-76. Changing hand, Step 7.

4. With a factory-supplied wrench or long-nosed pliers, straighten the tabs (Fig. 13-96).

5. Compress the slide with a screwdriver and pull the assembly out (Fig. 13-97).

6. Rotate the spindle 180° to relocate the knob catch (Fig. 13-98).

7. Rotate the slide assembly 180°. With the slide compressed, insert the lugs into slots on the mounting plate (Fig. 13-99).

8. Bend the tabs to secure the mounting plate (Fig. 13-100).

9. Install the linkage arm. (Fig. 13-101).

10. Install the lower bushing (Fig. 13-102).

11. Install the retaining ring (Fig. 13-103).

Fig. 13-75. Changing hand, Step 6.

Fig. 13-77. Changing hand, Step 8.

185

LINKAGE
ARM

INSIDE VIEW

Fig. 13-78. Changing hand, Step 9.

RETAINING
RING

OUTSIDE VIEW

Fig. 13-80. Changing hand, Step 11.

12. Install the driver bar (Fig. 13-104).

13. Install the bolt bar (Fig. 13-105).

14. Install the linkage bar (Fig. 13-106).

15. Install the knob driver (Fig. 13-107).

16. Install the linkage-bar plate (Fig. 13-108).

17. With the upper bushing slot in the horizontal position, install the cylinder driver (Fig. 13-109).

18. Replace the mounting-plate cover (Fig. 13-110).

19. Install the inside retaining ring (Fig. 13-111).

20. Turn the slot in the G85 unit to face the same direction as the slide (Fig. 13-112).

21. Turn the deadlatch 180° (Fig. 13-113).

22. Check your work (Fig. 13-114).

Attaching and Removing Knobs:

Figure 13-115 illustrates two of the most frequently

LOWER
BUSHING
(SLOT MUST
BE VERTICAL)

INSIDE VIEW

Fig. 13-79. Changing hand, Step 10.

DRIVER
BAR
(MUST BE
HORIZONTAL)

INSIDE VIEW

Fig. 13-81. Changing hand, Step 12.

186

Fig. 13-82. Changing hand. Step 13.

KNOB DRIVER

Fig. 13-84. Changing hand, Step 15.

encountered methods of securing door knobs. The knob may be threaded over the spindle and held by a set screw, or it may be pinned to the spindle by a screw that passes through the knob and spindle. Another approach is to secure the knob with a retaining lug that extends into a hole in the knob shank. The inside knob can be removed at any time by depressing the retainer; the outside knob can be removed only when the lock is open (Fig. 13-116).

UPDATING A LOCKSET

The following instructions concern the replacement of a worn or outdated lockset with a new one. The re-

LINKAGE BAR

Fig. 13-83. Changing hand, Step 14.

LINKAGE BAR PLATE

Fig. 13-85. Changing hand, Step 16.

Fig. 13-86. Changing hand, Step 17.

Fig. 13-88. Changing hand, Step 19.

placement is a heavy-duty "G" series Schlage, one of the most secure entranceway locksets made. The work is not difficult.

1. Remove the old lockset (Fig. 13-117).
2. Remove the latch (Fig. 13-118).
3. If a jig is available, use it as a guide for cutting the door; if not, use the template packaged with the lockset (Figs. 13-119 and 13-120).
4. Mortise the edge of the door to receive the strikeplate (Fig. 13-121).
5. Install the double-locking latch and deadbolt (Fig. 13-122).
6. Install the inside lockface assembly (Fig. 13-123).
7. Install the internal mechanism.

Fig. 13-87. Changing hand, Step 18.

Fig. 13-89. Changing hand, Step 20.

Fig. 13-90. Changing hand, Step 21.

8. Couple the outside lockface assembly with the internal parts (Fig. 13-124).

9. Secure the inside lockface assembly with the screws provided (Fig. 13-125).

10. Snap the inside cover into place (Fig. 13-126). The lockset is now installed (Fig. 13-127).

STRIKEPLATES

Quality locksets are equipped with a deadlocking plunger on the latchbolt. Otherwise the bolt could be retracted with a knife blade or a strip of celluloid. Nevertheless, it is important to leave very little space

Fig. 13-91. Changing hand, Step 22.

between the door edge and the strikeplate. A bolt with a ½″ throw should have no more than ⅛″ visible between the door and jamb. If necessary, mount the strikeplate over a steel spacer. This moves the strikeplate closer to the bolt.

The length of the lip is also critical since a short lip will increase wear on the latchbolt and may frustrate the automatic door-close mechanism, leaving the door un-

Fig. 13-92. Assembled left-hand unit.

Fig. 13-93. Right-hand conversion, Step 1.

Fig. 13-95. Right-hand conversion, Step 3.

latched. Mortise strikeplates are mounted on the same vertical centerline as the bolt (or should be). Measure the distance from the centerline of the latchbolt to the edge of the jamb and add ⅛″ for flat strikeplates and ¼″ for curved types. Figure 13-128 illustrates a collection of Schlage strikeplates for reference.

ENTRANCE-DOOR SECURITY

Security is a function of the door, cylinder, bolt, and strikeplate.

☐ There is *no* security with a hollow wooden door.
☐ Master keying should be kept as simple as possible.

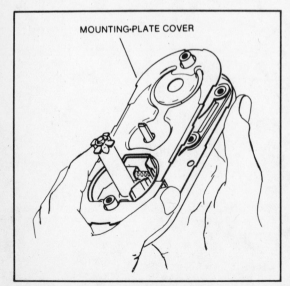

Fig. 13-94. Right-hand conversion, Step 2.

Fig. 13-96. Right-hand conversion, Step 4.

Fig. 13-97. Right-hand conversion, Step 5.

Fig. 13-99. Right-hand conversion, Step 7.

Each split pin increases the odds in favor of the lock picker.

☐ Resist demands for extensive cross keying. These demands originate with customers who insist upon single-key performance for executive keys.

☐ Security begins with proper key-control procedures.

☐ It is wise to supply extra change-key blanks during the installation phase of the system. Without readily available blanks, the convenient thing to do would be to cut duplicate change keys from master key blanks. Security could be compromised since the change key could fit other locks.

☐ Double-cylinder locks should be used wherever possible.

Fig. 13-98. Right-hand conversion, Step 6.

Fig. 13-100. Right-hand conversion, Step 8.

Fig. 13-101. Right-hand conversion, Step 9.

Fig. 13-103. Right-hand conversion, Step 11.

☐ Automatic deadlatches should be specified for all locksets without a deadbolt function. Otherwise the bolt could be retracted by "loiding," that is, by slipping a flat object between the bolt and the strike.

☐ Use the longest bolts available to frustrate attempts to gain entry by spreading the door frame. A strikeplate cover can also have the same function.

☐ Shield the bolts with armored inserts.

☐ Reinforced strikeplate mounts, particularly with metal door frames, add security by increasing the area of contact between the strikeplate and frame.

Sentry Door Lock Guards

Integral to improved lock security for home, apart-

Fig. 13-102. Right-hand conversion, Step 10.

Fig. 13-104. Right-hand conversion, Step 12.

Fig. 13-105. Right-hand conversion, Step 13.

Fig. 13-107. Right-hand conversion, Step 15.

ment, or business doors is the choice of the door strike to be used. All locks come with a strike that is especially configured for the lock, but you or your customer may wish to have a better, improved strike that will provide a greater amount of security. The Sentry Door Lock Guards Company, of Dania, Florida, has developed such a

line of strike units that will dramatically improve the security posture of individual door units. The designer and manufacturer of these unique lock guards guarantees that they are jimmyproof.

Perhaps you've never heard of this company or its president Frank Sushan, but many people and companies

Fig. 13-106. Right-hand conversion, Step 14.

Fig. 13-108. Right-hand conversion, Step 16.

Fig. 13-109. Right-hand conversion, Step 17.

INSIDE VIEW

Fig. 13-111. Right-hand conversion, Step 19.

have. As the inventor of these unique, impressive, and effective Lock Guard units, he has been written up in several major newspaper articles. In addition, numerous police departments across the nation now demonstrate the Sentry devices as part of local crime prevention lectures and demonstrations. Further, the National Crime Prevention Institute has approved and endorsed the installation of these products.

Figures 13-129 through 13-133 illustrate the various lock guards available.

LG-1 is a maximum security attachment produced for the do-it-yourselfer who may have no metal cutting tools or experience in security applications. You just remove the present striker and replace it with the fitted unit. The lock guard attachment is installed around the latch faceplate on the door, when such doors open out.

Fig. 13-110. Right-hand conversion, Step 18.

OUTSIDE VIEW

Fig. 13-112. Right-hand conversion, Step 20.

Fig. 13-113. Right-hand conversion, Step 21.

LG-2 is a single plate that is used with an existing striker plate for doors that open out. LG-2A is identical to the LG-2, with the exception of the closing section under the "T." The WDG-1 cross-section provides the extrusion view of the product.

The LG-3 is a single interlocking door plate for use on inswinging doors that are hung on a metal jamb door frame. All that is required is a slot to be cut into the metal door stop to receive the tongue of this interlocking door plate. Any door protected by this unit does not require locking with a key from the outside when leaving and the door is closed. The lock cannot be loided, jimmied, or jacked apart. Figure 13-133 shows a cross-section of the installation.

LINKAGE-
BAR PLATE

SLOT IN UPPER
BUSHING
HORIZONTAL

BOLT BAR

SLOT IN LOWER
BUSHING
VERTICAL

DRIVER BAR
HORIZONTAL

"R" SIDE OF UNIT

DEPRESS CATCH
AND INSTALL
OUTSIDE PLATE

Fig. 13-114. Right-hand conversion, Step 22.

A

B

Fig. 13-115. Securing door knobs.

The LG-4 is used similar to the LG-3, and when used alone the difference is that a slot must be made in a wood door jamb to receive the tongue of the unit.

The LG-5 is a two-piece unit for installation on doors that open inward only, and which are of wood. As indicated for the LG-3 and LG-4, such protected locks cannot be jimmied or jacked open. The doors and jambs will remain together and will not separate.

The LG-5A is a replacement safety striker plate for use on a wood jamb door frame in which the striker plate

Fig. 13-116. Removing knob.

Fig. 13-117. Lockset update, Step 1.

Fig. 13-118. Lockset update, Step 2.

Fig. 13-119. Lockset update, Step 3.

Fig. 13-120. Lockset update, Step 3.

Fig. 13-121. Lockset update, Step 4.

Fig. 13-122. Lockset update, Step 5.

Fig. 13-124. Lockset update, Step 8.

Fig. 13-123. Lockset update, Step 6.

Fig. 13-125. Lockset update, Step 9.

Fig. 13-126. Lockset update, Step 10.

extra measure of security for door locks. For old and new doors alike, these units are providing the ultimate security protection for the lock and also for the area immediately around the locking unit.

M.A.G. Engineering and Mfg. Co. has spent many years perfecting these type of units to provide ultimate security to the users of their products. As a locksmith, you cannot afford to be ill-prepared to meet customer needs when it comes to security applications concerning the locks you install. Because of the ever-increasing need for upgraded security in today's society, these products are widely used and beneficial to all. Figure 13-134 shows just one type of the Install-A-Lock lock changing kits available.

These kits can be applied to cylindrical locks, deadlocks, deadbolt and key-in-knob locksets, and also for mortise locksets (Fig. 13-135, 13-136). The lock changing kit enables you to install a lock on any door within minutes. They are made of heavy gauge stainless steel or brass, and the assembled unit brings the ultimate protection to new or old doors; it also covers damaged portions on burglarized doors. Even worn, old, mistreated doors can take on a new glow when graced with an Install-A-Lock kit. Designed to cover jimmy marks, splits and other

screws will not hold the plates onto the jambs, making the purchase of another new lock unit unnecessary. The LG-5B is a replacement safety striker plate used for the same reason as the LG-5A. This unit provides additional protection where entrance is gained by driving a tool through the molding behind the latchbolt, to force the bolt back to gain unauthorized entrance. The right angle tongue of the strike plate is extended under the mold for added protection.

LG-6 is an enlarged model of the LG-2 unit, for use on any old or new mortise lock application.

The WDG-1 and WDG-2 are seven foot blank extrusions that are weather strips as well as door and lock guards. They require that all holes will be drilled when the unit is installed. The length ensures that it will be long enough to fit any standard door frame, requiring only slight cutting down for perfect fits. Cross sections of these units are provided at Fig. 13-133.

As a point of customer reference, these lock guard units are much stronger and larger than those being used in other applications today. No stronger products are commercially available anywhere.

M.A.G. Install-A-Lock

Lock changing and door reinforcement kits provide a

Fig. 13-127. Lockset update, Step 10.

199

10-015

**½" Rabbeted
Box Strike**
For "C": 10-009; For
"D": 10-015
Size: 2¾" x 1⅛" x ⅛"
Lips: ⅞" through 3"
    (⅛" increments)
Strike shield may be ordered separately.
Specify finish.

10-003

**Standard ½" Rabbeted
Box Strike**

For: "A", "B", "E"
Size: 2¾" x 1⅛" x ³⁄₃₂"
Lips: ⅞" through 4"
    (⅛" increments)
Strike shield may be ordered separately.
Specify finish.

10-055

Box Strike

For: B400
Size: 2¾" x 1⅛" x ⅛"
For: B100, B200
10-033
Size: 2¾" x 1⅛" x ³⁄₃₂"

10-016

**Deep Box Strike (¾"
Throw) for Fire Door
Latchbolt**
For "C": 10-010, For
"D": 10-016
Size: 2¾" x 1⅛" x ⅛"
Lips: ⅞" through 3"
    (⅛" increments)

10-039

**Protected Back Box
Strike**
For: "A", "B", "D", "E"
10-038 — Lip: ¹⁵⁄₁₆" for
1⅜" doors
10-039 — Lip: 1¼" for
1¾" doors
Provides added latchbolt
protection on pairs of
doors.

10-040

Cast Open Back Strike

For: "A", "B", "D", "E"
Size: 2¾" x 1⅛" x ⅛"
Lip: 1" for 1¾" doors
Prevents latchbolt damage on pairs of doors.

10-006

Raised Lip Box Strike
SIZE: 2¾" x 1⅛" x
3/32" thick
LIP: 1¼"

10-004

**¼" Radius Rounded
Corner Box Strike**

For: "A", "B", "E"
Sizes 2¾" x 1⅛" x ³⁄₃₂"
Lips: ⅞" through 4"
    (⅛" increments)

10-027

**¼" Radius Rounded
Corner Strike**

For: "A" & "F"
Size: 2¼" High
Lip: 1¼" Full

Fig. 13-128. Schlage strikeplates.

defacements around old locks, it also provides new strength in the locking area. With the latch plate secured metal-to-metal—*not* wood—the completed installation is extremely solid and secure.

The one-piece construction provides a great savings in labor time. The doors are then protected, after only a few minutes installation time, right up to the leading edge on both sides of the door.

The security cost is economical and comes in a variety of sizes to meet varying requirements. The cylindrical lock kit will accept locksets with a 2" to 2⅛" bore, whereas the deadbolt and key-in-knob allow for two backsets of 2⅜" or 2⅜". Simply put, these are the finest remodeling kits available.

Fig. 13-129. The LG-1 and LG-2 units for door units opening outward. (courtesy Sentry Door Lock Guards Co.)

**Installation Instructions—
Deadlock or Night Latch:**

1. Remove the old lock and all screws from the door.

2. Bore holes for the new lock, if necessary, using the template furnished with the lockset to be installed in the door.

3. Chisel out the front edge of the door 1¼″ wide by 2⅝″ by ½″ deep in the center of the latch hole.

Fig. 13-130. LG-3 unit is a single, interlocking plate for in-swinging doors hung in a metal frame. (courtesy Sentry Door Lock Guards Co.)

Fig. 13-131. The LG-4 and LG-5 door guard units. (courtesy Sentry Door Lock Guards Co.)

Fig. 13-132. LG-5A and LG-6 lock guards. (courtesy Sentry Door Lock Guards Co.)

4. Slide the Install-A-Lock unit and position the large hole in the unit to line up with the hole in the door for the lockset.

5. After the unit is in position, install the lock in the door, using the 8-32 machine screws that come with the kit. These will secure the latch to the unit.

6. Install the four screws and washers to secure

Fig. 13-133. These blank extrusions (top view) are weatherstrips as well as door and lock guards. (courtesy Sentry Door Lock Guards Co.)

the unit to the door. Be sure that the face of the unit is flush with the door face.

7. Install the striker in the door jamb.

For mortise locks, if it is a new installation, follow the manufacturer's installation instructions, then follow the below instructions for the Install-A-Lock for mortise locks. For preparing a door in which a mortise lock is currently mounted, follow the instructions below only.

1. Remove the lock from the door.

2. Cut the door edge ⅝″ deep and 8½″ high centered on the lock.

3. Install two screw fasteners into the recessed area of the Install-A-Lock with the lip facing down.

4. To determine which side is long, place the unit on a flat surface. The unit will tilt slightly in the direction of the short side. Place the long side on the high edge of the door bevel.

5. Install the lock in the unit with the cylinder at the top.

Fig. 13-134. Install-A-Lock lock changing kit component parts. (courtesy M.A.G. Engineering and Mfg., Inc.)

Fig. 13-135. Install-A-Lock with a key-in-knob unit, plus a deadbolt locking unit added for extra security. (courtesy M.A.G. Engineering and Mfg., Inc.)

Fig. 13-136. Install-A-Lock for a mortise lockset. (courtesy M.A.G. Engineering and Mfg., Inc.)

Fig. 13-137. Door section with the old lock and latch removed. (courtesy M.A.G. Engineering and Mfg., Inc.)

Fig. 13-139. Line up the Adjust-A-Lock unit when fitting it to the door. (courtesy M.A.G. Engineering and Mfg., Inc.)

(For Sectional Trim)

6. Select the proper Install-A-Lock insert and place it between the door and the underside of the Install-A-Lock with the raised area out, fitting it on the large rectangular hole.

7. Install the cylinder and knobs to hold the Install-A-Lock unit in place.

8. Check to see that the lock matches the strike correctly.

9. Be sure the Install-A-Lock unit is tight against the door.

(For Escutcheon Trim)

6. Place the unit on the door.

7. Install the cylinder and knob assembly to hold the Install-A-Lock unit in place. *Do not* install the escutcheon retaining screws as yet.

8. Check to see that the lock matches the strike correctly.

9. Be sure the unit is tight against the edge of the door.

10. Mark and drill the top mounting hole closest to the edge of the door, using a 13/32″ drill, and install through bolt. Repeat the same procedures on the bottom outside hole closest to the door edge; tighten bolt securely. Drill and install the remaining three through bolts, installing the center bolt first.

11. For escutcheon, now drill and install the escutcheon screws as supplied by the lockset manufacturer.

For double-lock units, the following instructions should be followed:

1. Remove the lock, latch, and all screws (Fig. 13-137).

Fig. 13-138. Two mortise cuts are required in the door edge in addition to the holes required. (courtesy M.A.G. Engineering and Mfg., Inc.)

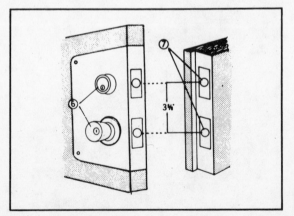

Fig. 13-140. Install the locksets and strikeplates. (courtesy M.A.G. Engineering and Mfg., Inc.)

Fig. 13-141. Install the strikes and mounting screws to complete the installation. (courtesy M.A.G. Engineering and Mfg., Inc.)

2. See Fig. 13-138. Mortise out the front edge of the door: (1) 2⅝″ high by ½″ deep in the center of the lock hole (2).

Using the template furnished with the deadbolt lockset, you mark and drill holes (3) and (4) 3⅝″ from the center of the bottom hole.

3. Slide the unit (5) over the door and position both holes to line up with the holes in the door (Fig. 13-139).

4. Install both locksets (6) (before installing the mounting screws). Install both striker plates (7) 3⅝″ between centers. The strike plate installation should be per the specific instructions for the installed lockset (Fig. 13-140).

5. Close the door and check to be sure that both the latch and deadbolt (8) enter the strikes (7). If they do not enter, move the Install-A-Lock assembly slightly up or down until the latches enter (Fig. 13-141).

6. Install the mounting screws (9) with finishing washers, being sure that the unit is flush with the door.

EXIT ALARM LOCKS AND PANIC BAR DEADLOCKS

A great concern of businesses that provide for locksmithing sales and servicing is the alarmed and deadbolted exit door lock. These units are used in the main area of controlling, monitoring, or warning of access (either inward or outward) through various doors and also into or out of areas that should be controlled and access limited. One-way (emergency) doors are one example. Controlled access areas, such as high-cost storeroom

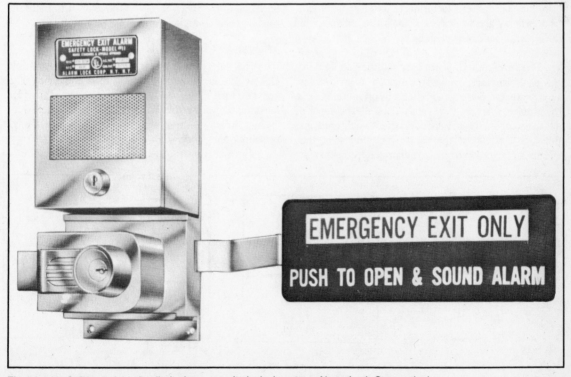

Fig. 13-142. Safetygard 11 deadbolt alarm security lock. (courtesy Alarm Lock Corporation)

Fig. 13-143. Safegard 70 Panic Bar alarm deadlock. (courtesy Alarm Lock Corporation)

MODEL 70R
OUTLINE DIMENSIONS
ALL DIMENSIONS ARE IN INCHES

KEEPER

CORRESPONDS TO WIDTH OF DOOR

Fig. 13-144. Safegard 70 specifications. (courtesy Alarm Lock Corporation)

stockage, back rooms of a small watch or jewelry store, are another example. For a restricted type area (again a watch or jewelry store, or more importantly, any business that has large amounts of money on hand or the drug storage area in a hospital) these lock alarm units are a great convenience and additional security.

The Alarm Lock Corporation, which specializes in these devices, has a long list of satisfied customers, including American industry, the U.S. government, many major chain and department stores, large food markets, major discount centers, factories, hospitals, post offices, many primary and secondary schools and colleges, and finally, small and medium sized businesses in your own city.

The deadbolt exit alarm lock and the double deadlock panic bar lock evolved to meet a major problem among various types of government and private concerns. The problem was that it was (and still is) illegal to lock or bolt an exit or emergency door because it would endanger life. Also, as long as the doors were unlocked, they became an invitation to the unauthorized intrusion from

the outside by undesirable elements of society. These intrusions led to the theft of many millions of dollars every year from the concerns that were illegally entered.

The solution to the problem was a locking device that allowed for emergency exit, sounding an alarm when the door was opened from the inside or out, but at the same time ensured that nobody could enter the area anytime they wanted. Such items as the Safetygard 11 Deadbolted Alarmed Security Lock (Fig. 13-142) and the Safegard 70 Panic Alarm Deadlock (Fig. 13-143) were only two of the items that were developed to meet the pressing needs of government and private business.

Safetygard 11 Exit Alarm Lock

The safety alarm lock provides secure full-time locking of emergency exit doors while complying with standard safety, fire, and local building codes. It provides maximum day and night security and a powerful deterrent against pilferage—from within or without—and always remains panic-proof.

The safety alarm lock provides instant automatic

Fig. 13-145. Modes of door operation with the Safegard 70 installed. (courtesy Alarm Lock Corporation)

Fig. 13-146. The MS1850A deadlock for narrow stile doors.(courtesy Adams Rite Mfg. Co.)

exit in case of an emergency. It combines the best features of panic devices, plus a deadbolt, plus a warning system. The first person reaching the door releases the lock with a slight pressure on the emergency clapper plate, which automatically releases the lock, opening the door and simultaneously sounding the alarm that the door has been opened.

Alarm Lock Operation. The complete lock mechanism retracts from the edge of the door when the clapper arm is depressed, which withdraws the deadbolt and sounds the powerful twin horn alarm. The alarm continues until the lock is reset to the original position with the appropriate key. The alarm can only be bypassed by an individual authorized to carry and use the bypass key. The mounting plate is already predrilled when the unit is received from the factory, so it is easy mounted on all

BACKSET	"A"
7/8" 22,2	1.57 39,9
31/32" 24,6	1.63 41,4
1-1/8" 28,6	1.78 45,2
1-1/2" 38,1	2.25 57,1

Fig. 13-147. Critical lock dimensions for the MS1850A deadlock. (courtesy Adams Rite Mfg. Co.)

Fig. 13-148. MS1850 A faceplate variations. (courtesy Adams Rite Mfg. Co.)

styles of doors. The lock is furnace mounted and requires no mortising.

Keying. The safety alarm lock unit is received without the cylinders. (This provides a means for you to sell the cylinder along with the unit, possibly keying the cylinder to meet a certain level of master key operation in medium and large size business and manufacturing complexes.) The safety alarm lock will accommodate all standard rim cylinders. This permits the lock to be integrated into an existing or proposed master key system. If special keying or rekeying is not required, the unit should be factory ordered with Alarm Lock's own cylinders; the increased cost is slight, but well worth it.

The lock may be key operated from the outside of the door by mounting an extra rim cylinder through the door. Outside key control cam and cover tamper alarm switch are built into every lock. The outside cylinder, though, must have a flat horizontal tailpiece. The overall length of the cylinder and tailpiece must be a minimum of 3½″ in length to suit a 1¾″ door.

Safegard 70 Panic Alarm Deadlock

This is the world's only available double deadlock panic lock, and it is the best. It provides for anti-pilferage alarm security from within and deadlock security against intrusion from the outside, plus it meets all requirements as an emergency exit device.

With deadbolt locking, the Safegard 70 is designed to deter the unauthorized use of the emergency and fire exit doors. Instant emergency panic exiting is provided, though, as authorized by law. An added advantage is that remote electrical control for this unit is available.

The unit can be operated in four different modes:

1. Maximum Security Locking—Once the door is closed, the bolt is projected by use of the key, firmly

See Page SW-100 for Lock Mounting Methods

Fig. 13-149. Lock and cylinder preparation, stile preparation and critical cam dimensions for the MS1850A. (courtesy Adams Rite Mfg. Co.)

Fig. 13-150. MS1890 Maximum Security Latch /Lock unit, with an armored strike and a lever handle. (courtesy Adams Rite Mfg. Co.)

securing the door by deadbolting it. The crossbar must be depressed to release the deadbolt and deadlatch to open the door. In doing this, the alarm is activated and remains so until the door is closed and relocked by the authorized relocking key.

2. Outside Key and Pull Access—By adding an outside cylinder and door pull, you have another mode of access. The deadbolt is retracted by the key and the deadlatch is released by the pull to open the door. Entry remains unrestricted from both sides until the deadbolt is relocked by the key from either the inside or the outside.

3. Outside Key Only Access—By adding the outside cylinder, in this mode the key must retract the deadbolt and release the deadlatch momentarily to open the door. The door relatches with each closing; the key is required for each re-entry. To relock the door and rearm the alarm, the deadbolt must be projected by use of the key.

Fig. 13-151. Latch /lock specifications. (courtesy Adams Rite Mfg. Co.)

INSTALLATION

(CUTOUT SAME AS SERIES MS1850A DEADLOCK PLUS HOLES FOR HANDLE MOUNT.)

CUTOUT TOLERANCE:
+ .015
— .000
+0,38
—0,0

CYL. CENTERLINE

.245
6,22

.694
17,6

4.885
124,0

5.375
136,5

.06
1,5

3.385
85,97

.750
19

.625
15,8

.156R
3,9
(4 PLCS.)

1.260
32,0

4.875
123,8

STRIKE

JAMB PREPARATION

(COVERS ANY PREVIOUS CUTOUT FOR M.S.®DEADLOCK)

MAXIMUM GAP, DOOR TO JAMB: 1/8"
3,1

1.010
25,6

.156 R
3,9
(4 PLCS.)

BACKSET

1.594
40,4

1.25
31,7
DIA.

6.885
174,8

.068
1,72

5.556
141,1

1.50
38,11

3.00
76,2

.438
11,1
DIA.

STILE PREPARATION

2 PLCS:
PADDLE:
.265 DIA.
6,7

LEVER:
TAP 10-32

LATCHBOLT HANDING EASILY REVERSED IN FIELD

LATCH/LOCK, CYLINDERS AND PADDLE INSTALLATION

Fig. 13-152. Metal stile jamb and paddle /cylinder installation procedures in an exploded view. (courtesy Adams Rite Mfg. Co.)

4. Bar Dogging Function—This provides for free passage from both sides of the door.

The Safegard 70 dimensions are shown at Fig. 13-144. Modes of operation for single doors, pairs of doors, and pairs of doors with center mullions are shown at Fig. 13-145.

NARROW STILE GLASS DOORS

Narrow stile glass doors have evolved the need for a more secure type of locking device than is used in other applications. The architectural trend for such types of doors will continue for quite a number of years to come. As an individual concerned with security, you must be extremely aware of what is available to meet a variety of locksmithing situations, both in the home and the general business environment.

First, you must realize that because of this architectural trend, the aluminum used in narrow stile glass doors is shrunk in width and material distance from that of the wooden door that would have a large piece of glass mounted in it. Because the aluminum is narrower and of less thickness, it naturally becomes slightly more flexible. At the same time, of course, the space to house the lock also becomes smaller. To retract a long bolt in the conventional matter is impossible since there is no place for it to go. The conventional lockbolt in an aluminum frame of a glass door has a ½″ to ⅝″ throw with an effective penetration of the strike, in some instances, of as little as ¼″. This door can be sprung open by just about anyone with a couple of screwdrivers or, if the frame is weak or worn, with a well-directed frontal attack by a moving body (the intruder). This highly attractive situation means that the lock installed must be able to stop the hammer and bar and general forced entry.

To an extent, the maximum security considerations for glass doors that *slide* must also be considered. These doors have spawned some misconceptions, some fostered by locksmiths who really don't know what they are talking about because they haven't studied the doors, their locks, and the other applications.

This confusion has resulted in some of these doors being installed still lacking the most basic elements of protection that any door should have. Chief among these are a *real* lock—a secure lock, a lock that can be operated by a key if desired and also one in that the key is compatible with the system of other locks in the same building.

Convenience *and* security are both necessities to the owner/user of the lock, and this is where we now consider the specialized locks that offer the maximum security and protection available for these types of doors. Only one company specializes in locks and locking devices for narrow stile doors, and that is the Adams Rite Manufacturing Company, of California. The next several pages will cover their products.

Deadlocks

The MS1850A deadlock (Fig. 13-146) is the first in a series of specialized locks developed especially with the narrow stile door in mind. This lock uses a high bolt of laminated steel, nearly three inches long and actuated by an uncomplicated pivot mechanism. This lock has made the Maximum Security Deadlock the standard for the entire narrow stile door industry. The length of the bolt

Fig. 13-153. Latch bolt holdback instruction details. (courtesy Adams Rite Mfg. Co.)

provides the maximum security for any single leaf door, even a tall and flexible one. It is also excellent for installation applications where the gap between the door and the jamb is greater than it should be.

The lock case measures 1″ × 5 3/16″ × the desired depth. The case depth varies with the backset (Fig. 13-147). The cylinder backset can be ⅞″, 31/32″, 1⅛″ or 1½″. Remember that in measuring the backset, it is from the centerline of the faceplate to the centerline of the cylinder.

The bolt is ⅝″ × 1⅜″ × 2⅞″ with a 1⅜″ throw. It is constructed of a fly ply laminated steel, the center ply having an Alumina-Ceramic core to defeat any hacksaw attack, including rod-type "super" hacksaws.

The faceplate for the MS1850A comes in five variations, as illustrated in Fig. 13-148. Figure 13-149 shows the dimensions (in inches and millimeters), illustrates the stile preparation for the lock, lock and cylinder installation, and information about the cylinder cam.

Latch Locks

The MS+1890 latch/lock (Fig. 13-150) is a deadlocked Maximum Security unit offering the highest type of protection after-hours for the businessman or homeowner in addition to traffic control convenience for management (in businesses) during the business day.

A typical installation in a bank or a store could require any of three modes of door control: (1) both the lock and latchbolts retracted for unrestricted entry and exit during business hours; (2) handle-operated latch for exit-only traffic; and (3) maximum security hookbolt for overnight lockup.

This has the operational advantage not necessarily provided by others which may seem similar. It requires a 360° turn of the key to throw or retract the hook-shaped deadbolt. When unlocked, a 120° further turn of the key will also retract the latchbolt as well. When using the door handle that can also be obtained for the lock, the handle retracts the spring-loaded latchbolt only.

Figure 13-151 shows the overall dimensions of the lockset. Figure 13-152 has the stile preparation, strike location and dimensions necessary, and the latch/lock, cylinders and paddle installation. Latchbolt holdback information is at Fig. 13-153.

Threshold Bolts

Along with a lock installation, you may be required to install a corresponding threshold bolt mechanism also in a narrow stile door (Fig. 13-154). The use of the

Fig. 13-154. Adams Rite threshold bolt (left) and header bolt (right); unit has the MS1850A maximum security lock installed. (courtesy Adams Rite Mfg. Co.)

Fig. 13-155. Threshold bolt and header critical dimensions. (courtesy Adams Rite Mfg. Co.)

threshold bolt allows maximum security by the turn of a single operating key by the user for pairs of doors. With the simultaneous dropping of a hardened steel hexbolt into the threshold and pivoting of the doors' stile, the two-point lock secures the entire double door entrance-way. The threshold bolt is harnessed to the rear of the pivoted bolt. The standard threshold bolt rod is sufficient for a cylinder height up to 53 7/17″ maximum. It is fully threaded and can be cut off to allow for low cylinder heights.

A full turn of the key (or a thumbturn if one is installed on the interior side of the door) throws the counterbalanced bolt into the opposite door and the drop bolt into the threshold. The key can only be removed when the bolts are in a positive locked or unlocked position—not halfway! Figure 13-155 shows the dimensions, Fig. 13-156 the installation of the threshold bolt, and Fig. 13-157 illustrates the header bolt installation.

Narrow Stile Door Deadlock Strikes

Three basic strikes are available. The majority of Adams Rite deadlocks are installed in metal construction

ANY BASIC M.S. DEAD-LOCK

INSTALL ADAPTOR ARM IN LOCK FROM REAR WITH BOLT RETRACTED AS SHOWN. DRIVE PIN THROUGH SLOTS IN LOCK SIDE PLATES PASSING THROUGH HOLE IN THE ADAPTOR ARM.

ROD/BOLT LENGTH MUST BE APPROXIMATELY 7" LESS THAN CYLINDER HEIGHT.

178

CUT OFF TO SUIT LOW CYLINDER HEIGHTS.

STILE PREPARATION

SEE BASIC M.S. DEADLOCKS MS1850A MS1850 OR MS1850ANSI

.562
14,3

.69
17,5

.63
16,0

¢ BOLT

.91
23,1

DRILL & C'SK FOR #10-32 FLAT HEAD SCREWS ON INTERIOR FACE OF STILE.

.500
12,7
DIA. HOLE

THRESHOLD

STRIKE

HEADER PREPARATION FOR STRIKE

#6-32 TAP (2) HOLES

.25
6,3

1.00
25,4

1.500
38,1

1.75
44,4

STRIKE

.38
9,6

TOP OF DOOR

1.500
38,1

STILE PREPARATION

STRIKE FOR M.S. PIVOTED BOLT PLUS HOLES FOR 4 SCREWS:
.199 DIA. HOLE
5,05
.396
C'SK 82° × 10,06
DIA. (4 PLCS.)

BREAK OFF TAB AFTER INSERTING SCREWS

DRILL & C'SK FOR #10 FLAT HEAD SCREWS (4)

.500
12,7

.50
12,7

STRIKE CUTOUT FOR M.S. LOCK

Fig. 13-156. Stile preparation and installation for the threshold bolt. (courtesy Adams Rite Mfg. Co.)

Fig. 13-157. Installation details for the header bolt. (courtesy Adams Rite Mfg. Co.)

214

where the strike cutout can be simply a slot. For aesthetic reasons, or, in the case of certain locks, for added security, specific strikes should be used.

The trim strike (Fig. 13-158) is a simple strike plate that can be surface-mounted on a wood or metal jamb or mortised flush. It uses #10 flathead machine screws and/or two #10 wood screws. It is made of aluminum.

The box strike has a dust box added (Fig. 13-159). It is customarily used only for wood construction where the dust box prevents chips, sawdust, and other debris from entering the strike.

The armored strike (Fig. 13-160) has the basic trim plate, but this is backed up by a massive steel doubler designed to prevent the method of forced entry known as "jamb peeling." It fits within aluminum or other hollow jamb sections; with a trim face flush, the steel is completely hidden.

Figure 13-161 shows mounting methods for narrow stile aluminum doors. The three most prevalent methods of mounting locks and latches within the hollow tube stile of a door are basically easy.

Fig. 13-159. Box strike, customarily for wood frames where the dust box permits debris from entering the strike. Can also be used on metal door frames. (courtesy Adams Rite Mfg. Co.)

Fig. 13-158. Simple trim strike plate for deadlocks in a metal stile door unit. (courtesy Adams Rite Mfg. Co.)

Fig. 13-160. Armored strike unit, essentially used to prevent jamb peeling types of forced entry. (courtesy Adams Rite Mfg. Co.)

METHOD "A"

RESILIENT WASHER

4104 BRIDGE

LOCK MOUNTING PLATE

MOUNTING SCREW

Method "A" spans the stile with a steel bridge at top and bottom of the lock. A simple handle (Adams Rite Installation Tool 4075) is used to position the bridge accurately in the stile while its two screws are tightened to form a "web" in the door. Resilient washers allow for minor adjustment to fit the lock flush in door.

METHOD "B"

TAPPED HOLE

TUBE SPACER

.25
6.35

MOUNTING SCREW

LOCK MOUNTING PLATE

Method "B" requires two tapped holes in the back web of the stile and tubular spacers of appropriate lengths. Some door manufacturers use a heavy coil spring in place of the tube to allow for minor adjustment.

METHOD "C"

FLANGE CLIP 4435

RESILIENT WASHER

LOCK MOUNTING PLATE

MOUNTING SCREW

Method "C" is used with doors whose stile extrusion shape contains a special inside channel to accept flange clips on which the lock is mounted. Several door manufacturers have incorporated this channel in their newest designs. It provides a convenient lock mounting means as well as adding overall strength to the stile.

Fig. 13-161. Three methods of mounting a lock in the narrow stile doors using the basic component parts of the Adams Rite installation kit. (courtesy Adams Rite Mfg. Co.)

Method A uses the steel bridge spanning the stile. A simple handle, the installation tool, is used to position the bridge accurately in the stile while its two screws are tightened to form a web in the door. Resilient washers allow for minor adjustment to fit the lock flush with the door.

In method B, two tapped holes in the back web of the stile and tubular spacers of the appropriate lengths are required. Some door manufacturers use a heavy coil spring in place of the tube to allow for minor adjustments.

Method C is for doors whose stile extrusion shape contains a special inside channel to accept flange clips on which the lock is mounted. Several door manufacturers have included this channel in their latest designs, so it is something you must be aware of and look for prior to installing the lock. It provides a convenient lock mounting means as well as adding overall strength to the stile. *Note:* Refer to the business chapter for the Adams Rite Locksmithing Installation Kit which discusses some of the specific tools necessary for the three methods of mounting locks.

Fig. 13-162. Cylinder pull set for both sides of a sliding door. (courtesy Adams Rite Mfg. Co.)

DIMENSIONS

INCHES
MILLIMETERS

10-32 TAP (4 HOLES)

.87
22,1

.43
10,9

2.91
73,9

2.55
64,8

1.97
50,0

.72
18,3

SPINDLE

1.25
31,7

.12
3,0

.12
3,0

STANDARD PULLS	
PULL NO.	CYLINDER INCLUDED:
4025	ADAMS RITE
4025-1	NONE (DUMMY)
4025-5	SCHLAGE 22-001
4025-6	WEISER KEYWAY

SPECIAL PULLS	
PULL	FURNISH CYLINDER:
4025-3	LOCKWOOD R-71, H-71, S-71
4025-4	SARGENT SIXLINE
4025-8	CORBIN 471AR
4025-9	RUSSWIN AR848
4025-10	DEXTER 9094
4029	BEST 5A5A2
4025-15	MEDECO 20-800

TAP 6-32
(2 HOLES)

SPINDLE

.975
24,76

1.950
49,53

2.28
57,9

1.22
31,0

.78
19,8

Fig. 13-163. Cylinder pull dimensions breakout of cylinders that can be used with the unit. (courtesy Adams Rite Mfg. Co.)

Fig. 13-164. Installation of the cylinder pull; two different pull variations shown. (courtesy Adams Rite Mfg. Co.)

Cylinder Pulls

The cylinder pulls (Fig. 13-162) are designed for certain deadlocks and latches. These surface-mounted pulls allow a sliding glass door to be keyed in to other types of doors which have a key-in-knob or mortise cylinder lock installed. Dimensions for the cylinder pull are at Fig. 13-163 and installation instructions shown at Fig. 13-164.

Lock Cylinder Guards

The standard mortise cylinder is made of brass and is literally the "soft spot" in the narrow stile door's security. Using special pliers, or other leverage devices, a burglar can tear the cylinder out of the door, leaving an opening through which he can operate the deadlock by hand. Against this problem has been developed the cylinder guard by Adams Rite (Fig. 13-165). It offers a three-way defense. First, the outer shield ring offers a poor purchase for either prying or twisting. Second, the ring is of hardened steel so that the combination of shape and hardness makes it virtually impossible to grip, even with sharpened tools. Third, in the event that a prying tool, such as a cold chisel, can be (if ever) driven into the stile behind the shield ring, the would-be burglar is obliged to

Fig. 13-165. Hardened steel free-swiveling security ring and retainer plate to provide protection against direct attack on the locking unit. (courtesy Adams Rite Mfg. Co.)

DIMENSIONS

INCHES
MILLIMETERS

STILE
WALL

1.312
33,32

2.437
61,90

.650
16,51

CYL.
LENGTH (INCL.
CAM) *

1.50
38,1
DIA.

1.625
41,27 DIA.

.710
18,03 MIN. CYL.,
CAM TO SHOULDER *

.43
10,9

Fig. 13-166. Cylinder guard dimensions and positioning when installed. (courtesy Adams Rite Mfg. Co.)

FACE "OPEN"
SIDE OF RETAINER
TOWARD MINIMUM
CLEARANCE DIRECTION
(TOWARD NOSE OF
DOOR ON
⅞" OR ³¹⁄₃₂" BACKSET LOCKS,
AWAY ON 1⅛" BACKSET).

NOTE:
GUARD MAY BE USED ON
OUTSIDE ONLY OR ON
BOTH OUTSIDE AND INSIDE
OF A DOUBLE-CYLINDER LOCK

RETAINER

SPACER(S)
TO FIT STILE
WALL IF
THINNER THAN
NORMAL
1/8" OR 1/4"
3,2   6,4

SECURITY RING
(HARDENED STEEL)

Fig. 13-167. Cylinder guard installation. (courtesy Adams Rite Mfg. Co.)

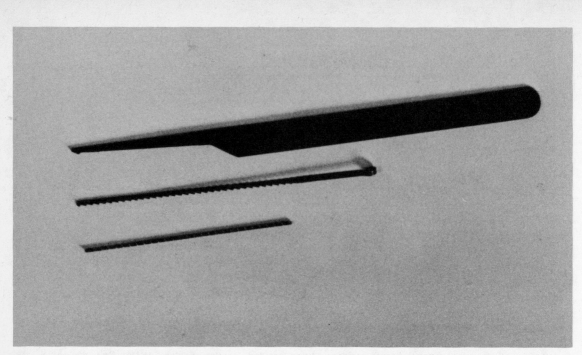

Fig. 13-168. Three homemade key extracters.

pull a heavy steel plate through the round hole in the metal door stile. This degree of leverage is far beyond that available with most hand tools.

The standard package includes the security ring which is of hardened steel (and free-swiveling when properly installed) and the steel retainer plate to permit the security ring to swivel. The latter is hardened to act as an anti-drill shield. The package also contains spacers for flush fitting in thin stile walls.

Figure 13-166 shows the cylinder guard dimensions.

Figure 13-167 provides installation information.

EXTRACTING BROKEN KEYS

Figure 13-168 illustrates three homemade extractor tools. The first type, made from a length of spring stock, is inserted with the hook down. The rounded tip of the tool nudges the pins up, and the hook fits into the most forward key cut. The second two are made from coping or fret saw blades. They are inserted so the teeth snag the edges or top of the key blade.

Chapter 14

Auxiliary Door Locks

The term *auxiliary* applied to door locks has two distinct meanings. It can mean a second lock acting in conjunction with a first and, hopefully, increasing the security from intruders. These locks are usually surface-mounted, in which case they are known as *rim locks or locking bodies*. Locks used on interior doors are also known as auxiliary locks. Most of these are mortise locks, i.e., they mount inside the door.

LOCKING BODIES

Locking bodies come in a variety of styles and shapes. A few of the options are shown in Fig. 14-1. The bolt may lock automatically when the door is closed. The two locking bodies at the top of Fig. 14-1 employ these self-locking latchbolts. Other locking bodies use dead-bolts that must be manipulated to lock. Either type may be worked only from the inside or have a keyway on the outside of the door.

Keyless locking bodies with latchbolts have one big disadvantage: they invite lockouts.

Construction

The basic mechanism is a case (or cover), a bolt, and a bolt-actuating mechanism. Latchbolts are spring-loaded and are checked by a safety catch. Engaging the safety from the inside puts the bolt and spring in check. The door can be shut without latching.

Locking bodies are adaptable to many different doors. Since they mount on the inside surface of the door, the bevel of the door edge is unimportant. But there are times when you will have to choose a particular locking body because of the configuration of the door frame. If at all possible, use the strike intended for this lock since strikes and locks work best as a matched team. If this is not possible, be certain that the strike you use works properly. Figure 14-2 illustrates two of the many strikes available.

Installation

Every locking body worthy of the name comes with installation instructions, but you will sometimes find yourself working "blind." In the absence of manufacturer's instructions, follow this procedure:

1. Position the locking body on the door and scribe its corners and locking body screw holes.
2. Remove the locking body and align the strike on the door frame.

Fig. 14-1. A collection of locking bodies.

3. Tentatively mark the strike position.

4. Assuming that an outside cylinder will be fitted, drill a hole through the door at the estimated point of contact of the tailpiece (the extension on the back of the cylinder) and the bolt-activating mechanism.

5. Measure the diameter of the cylinder and chuck up the appropriate bit or hole saw. Most cylinders are 1⅜″ in diameter.

6. Drill the hole from the outside using the first hole as a guide.

7. Break off the splinters and smooth the edges of the hole.

8. Mount the cylinder over the collar and attach it to the retaining plate on the inside of the door.

9. Hold the latch in position and check that the cylinder works properly. You may have to trim the tailpiece.

10. Check the screw holes against the locking body.

11. Start the screw holes with a small nail and drill 1/16″ pilot holes ⅓″ deep.

12. Mount the locking body and tighten the screws.

13. It may be necessary to chisel out part of the frame behind the strike. Some are surface-mounted.

The customer may want a set of instructions so he can do the work himself. In this case, the following instructions would be helpful:

Fig. 14-2. Two strikes by Corbin.

222

1. Drill a cylinder hole in the door 2⅜″ from the edge for doors opening inward, and 2½″ for doors opening out.

2. Install the cylinder. Make sure the pins are on top of the keyway.

3. Position the locking body even with the edge of the door and check that the tailpiece operates the latch when the key is used.

4. Attach the locking body.

5. Install the strike.

Fig. 14-4. A massive heavy-duty bolt deadlock. (courtesy Taylor Lock Company)

INSIDE DOOR LOCKSETS

The Corbin 7000 series is typical of the better locksets (Fig. 14-3). Designed for exterior or interior use, it has these features:

☐ Case—cast iron and lacquered.
☐ Deadbolt—forged brass with a ⅝″ throw. Hardened steel inserts are optional.
☐ Latchbolt—extruded brass with a ⅝ or ¾″ throw, depending upon application. Antifriction latchbolts are available.
☐ Latch holdback (safety catch)—optional.
☐ Hub—cold-forged bronze.
☐ Strike—wrought brass or bronze with the edge of the lip 1⅛″ from the centerline as standard.

Installation

Installation of Corbin and other locksets has been discussed in the previous chapter.

The Taylor 850 (Fig. 14-4) is a massive bolt rim deadbolt specifically designed for the extra measure of

Fig. 14-3. The Corbin 7000 series mortised lockset.

Fig. 14-5. Apartment lock. (All drawings in this series courtesy of the Emhart Corporation.)

Fig. 14-6. Communicating lock.

Fig. 14-9. Entrance or classroom lock.

Fig. 14-7. Dead locks. The lock may be controlled by a single key (A), two keys (B), or by a key and a turnpiece (C).

Fig. 14-10. Entrance or office locks may have an outside key (A), an outside key and a turnpiece (B) or inside and outside keys (C).

Fig. 14-8. Dormitory lock.

Fig. 14-11. Entrance or storeroom lock.

protection required on doors in a home or business. The bolt measures 1 11/16″ high by ⅝″ thick, which is twice the size for a conventional bolt in a locking unit of this type. In addition to a full one inch extension beyond the case, the bolt also possesses two hardened steel inserts which are intended to block sawing the bolt in half.

The strike, like the lock itself, is also massive in design and size and unlike more conventional strikes, uses five screws set on two face angles to provide for an optimum penetration and holding strength once applied.

From the inside, the deadlock is operated by a thumbturn, and from the exterior by a key. Normally, the

Fig. 14-12. Hotel room lock.

Fig. 14-14. A privacy lock.

Fig. 14-13. Passage or closet latch.

Fig. 14-15. Storeroom or closet latch.

unit is randomly keyed at the factory, but when several are required, they can be keyed alike (KA) or master keyed (MK), as circumstances warrant.

Like other units of this general type, lock installation is easy and takes an extremely short time to mount and put into operation.

VARIANTS

Indoor locksets have evolved according to function:

Apartment Lock. —The latchbolt is released by either knob, unless the outside knob is locked by the stop button or deadbolt. The stop button can be released manually at the button or by turning the inside knob. Turning the key retracts the deadbolt; turning the inside knob retracts the deadbolt and the latchbolt (Fig. 14-5).

Communicating Lock. —The latchbolt is released by either knob. A split deadbolt allows operation by either turnpiece (Fig. 14-6).

Deadlock. —There are three variations to the familiar deadlock: deadbolt operation by key from one side only (Fig. 14-7A); deadbolt operation by key from both sides (Fig. 14-7B); deadbolt operation by key on one side and turnpiece on the other (Fig. 14-7C).

Dormitory Lock. —The latchbolt is released by either knob, unless the outside knob is locked by the deadbolt. The deadbolt deadlocks the latchbolt. The deadbolt is activated by the inside knob (which also retracts the latchbolt and unlocks the outside knob) by turning the turnpiece or the key (Fig. 14-8).

Entrance or Classroom Lock. —The latchbolt is released by the knob on either side, unless the outside key locks the outside knob. In this mode the inside knob still opens the door (Fig. 14-9).

Entrance or Office Lock. —There are three variations. The first does not have a turnpiece (Fig. 14-10A). The latchbolt is released by the key from the outside and either knob, unless the outside knob is locked by the key. Alternately, this same type of lockset may be fitted with a turnpiece on the inside (Fig. 14-10B). In this case, the latchbolt is worked by the key on the outside and the knob on either side, unless the outside knob is locked by the key. The deadbolt is released by the inside turnpiece and the outside key. The third variation has a key on both

225

Fig. 14-16. Inside and outside trim. (courtesy Emhart Corporation)

sides (Fig. 14-10C) and works like the lock just described, except that the key, rather than the inside turnpiece, works the deadbolt.

Entrance or Storeroom Lock. —The latchbolt is activated by the knob on either side. The deadbolt operates by the outside key or the inside turnpiece (Fig. 14-11).

Hotel Room Lock. —The deadbolt is activated by the turnpiece on the inside. The latchbolt is retracted by the inside knob and by the guest, master, and grand master keys. The inside knob retracts the deadbolt and latchbolt simultaneously. Only the emergency key can release the deadbolt from the outside (Fig. 14-12).

Passage or Closet Latch. —Turning either knob releases the latchbolt (Fig. 14-13).

Privacy Lock. —The latchbolt is released by either knob. The deadbolt is activated by the inside turnpiece or by the inside knob (Fig. 14-14).

Storeroom Lock or Closet Lock. —The latchbolt is released by either knob. The deadbolt is activated by a key on the outside (Fig. 14-15).

Fig. 14-17. Turnpieces by Corbin.

Fig. 14-18. Knob and lever spindles by Corbin.

FOR WOOD DOORS

WASHER
128F36-8
FOR
1 ¾
DOOR

LOCK
MORTISE

ADAPTER
PLATE
189F09

FOR 1 3/4" DOOR
SPECIFY
FOR OTHER DOOR THICKNESS

(80034 SERIES)

VIEW OF OUTSIDE TRIM

PIN
228F54

SPINDLE

NYLON BUSHING

THIMBLE
TO SUIT
ROSE DESIGN

FILLER COVER
TO SUIT
KNOB DESIGN

OUTSIDE SHANK

OUTSIDE KNOB
AS CATALOGED

ADAPTER

ROSE AS
CATALOGED

SLEEVE SCALP

KNOB FASTENING RING

Fig. 14-19. Knob mechanism, exploded drawing.

SPINDLE STYLES AND KNOB SCREWS

STRAIGHT

432 . . . 3½"   433 . . . 4"   433A . . . 4½"

434   3¼"   435A   4"
435   3⅜"   436 . . . 4½"

SWIVEL

424 . . . 4½"

425 . . . 4½" x ⅜" Centers

426 . . . 4½"

427 . . . 4½" x ⅜" Centers

SPECIFICATIONS:
Thread: All threaded spindles, 20 threads to inch. Drilled and tapped holes, 9/24 thread.
Size: All spindles on this page, 9/32" square.
Centers: Center bushings where supplied, ⅜", are removable.
Packing: 1 Dozen per box.
Weight: Average weight, 1 lb. per dozen.

UNIVERSAL

437 . . . 3⅝"   438 . . . 4½"

439 . . . 4½"

440 . . . 4½" x ⅜" Centers

441 . . . 2½"

SPLIT

428 . . . 2½"

430 . . . 2½"

KNOB SCREWS

KSS
Set Screw
¼-28 x 9/32"

SKS
Side Knob
Screw
9-24 x ⅜"

KW
Knob Washer
⅜" I.D.

NOTE! All spindles shown on this page are machined to twenty threads per inch. In most cases, these spindles can substitute for spindles found in some comparable locks of another brand. There are locks, however, that use spindles with varying threads. Do not attempt to force spindles since this will damage threads.

CLOSET SPINDLES
(Brass Plated)

460   3¼" with rose

461   3¼" with rose

Fig. 14-20. Spindle styles and knob screws in demand by customers. Included here are individual unit dimensions allowing you to select correct spindle to fit specific lockset. (courtesy Taylor Lock Company)

HARDWARE

Figure 14-16 illustrates a sample of inside and outside door trim offered by one manufacturer. Figure 14-17 shows five turnpiece (thumb knob or turn knob) options from the hundreds that are available. Many options are purely cosmetic; others are mechanical parts that must be replaced with the exact equivalent. Prime examples are knob and lever spindles (Fig. 14-18). You would do well to stock a variety of these parts (Fig. 14-19).

Every locksmith, especially the beginning one, must have a variety of spindles and knob screws on hand, for repair, direct sales, or for specific job requirements. With so many different types available, you cannot survive without the variety immediately required. Figure 14-20

shows the technical specifications of the various types of spindles and knob screws.

These spindles are threaded at 20 threads to the inch. They have drilled and tapped holes at 9/24 thread and are 9/32" square. In most cases, these spindles can substitute for spindles found in other comparable locks of various brands. There are locks, however, that use spindles with varying threads. Do not attempt to force spindles, because this will damage the threads.

The set screws, knob washers, and closet spindles are quite often used for specific lock installations, and these sizes should be kept on hand, usually in the same storage box with or next to the spindles, because they are closely related and used together.

Chapter 15

Office Locks

Office locks represent an important and lucrative market and one that is easy to overlook. An important part of the security of every business depends upon office locks. A profusion of manufacturers are in this field, making thousands of different locks, each with particular features. But almost all of these locks fall into three groups: rotary-cam locks, sliding-bolt locks, and plunger locks.

ROTARY-CAM LOCKS

Simplest of office equipment locks, most rotary-cam locks employ disc tumblers, although you will occasionally run across one with pin tumblers. The cam is oval and rotates with the cylinder. Those with a 90° throw must be locked for key removal; 180° throw locks can be unkeyed in the locked or unlocked position (Fig. 15-1).

The lock body may be threaded (as shown in the illustration), or it may be secured by a retaining clip. The cam may be integral with the lock, or it can be secured by a screw and washer. A removable cam is an advantage since it allows the same lock body to be used in a variety of applications. Locks that turn 90° have a single tumbler chamber; locks that turn 180° employ a double chamber and may have a stopping pin, notch, or blocking member at the rear that limits rotation. In some cases it is possible

to change the rotation in the field. The lock may then be used in other applications.

SLIDING-BOLT LOCKS

Most of these locks are disc tumblers, although pin tumblers are not unknown. There is no cam; the bolt is grooved to accept a projection on the back of the plug (Fig. 15-2). The projection engages the groove and converts the rotary motion of the plug into reciprocating (to and fro) motion. The bolt is generally heavier and more durable than those of other office locks. This arrangement allows the key to rotate 180°; the key can be withdrawn in either the locked or unlocked position.

The most serious sliding-bolt lock is known as the T-bolt because of its shape (Fig. 15-3). These locks almost always employ a disc-tumbler cylinder with the bolt-actuating pin cast as part of the plug. The plug can generally be released with a probe wire.

If the key is not available and the bolt is thrown, you have three options. The first, and the best, is to pick the lock. Once the drawer is open, removing the retainer is no problem. Another approach is to drill through the drawer. Figure 15-4 shows the approximate drilling point. Drill no deeper than ⅞" to avoid damage to the lock

Fig. 15-1. A Medeco high-security cam lock.

SLIDING
BOLT

CTSK.
FOR
NO. 6
SCREWS

$1\frac{1}{4}$

$\frac{5}{8}$

$\frac{3}{16}$

$\frac{3}{4}$

$1\frac{7}{32}$

$1\frac{3}{4}$

$\frac{3}{16}$

$\frac{1}{4}$

$\frac{7}{8}$

$1\frac{5}{8}$

$\frac{7}{8}$

$1\frac{1}{8}$

SURFACE   HALF MORTISE

Fig. 15-2. A Corbin sliding-bolt lock, widely used on cabinets and drawers.

Fig. 15-3. A Corbin T-bolt lock, generally stronger than other sliding-bolt locks.

itself. Finally, you can drill out the plug, destroying the lock in the process.

PLUNGER LOCKS

Plunger locks are usually found on sliding cabinet doors. These locks are mounted on the outside door and the bolt, extending from the rear of the plug, engages a hole in the other door (Fig. 15-5). The doors are locked against each other so neither can be moved. A spring returns the plug to the open position when the key is turned. Although disc-tumbler types once were popular, pin-tumblers are almost universal today. Locksmiths new to the trade may never have seen a disc-plunger lock.

The plug is usually secured by a screw located at the side of the cylinder and set in a notch in the lock (Fig. 15-5). The screw serves a double purpose: It determines the throw of the bolt and aligns the plug with the cylinder.

The bolt is a projection from the rear of the plug unit. Without the key you can pick the lock or force the doors far enough apart to disengage the bolt. The latter method is the faster of the two.

FILING-CABINET LOCKS

Filing-cabinet locks are fairly well standardized and use a bolt extending from the top of the lock body (Fig. 15-6). The bolt manipulates the control bars that lock each drawer of the cabinet.

The bolt is spring-driven; unless the plug is turned, the bolt extends out of the locking body. In other words, the drawers are automatically locked.

The bolt may be round or rectangular in cross section. The lower face of the bolt—the part buried in the cylinder—is recessed for the spring. The plug cam works in a notch on the side of the bolt.

Fig. 15-4. Occasionally it is necessary to drill a desk lock for access to the retaining clip.   A 1/16" hole, ⅞" deep or less, is adequate.

Fig. 15-5. A Corbin plunger lock, intended to secure sliding doors.

Fig. 15-6. A filing-cabinet lock in cross section.

Fig. 15-7. A jimmy for open-keyway locks. The tool is made of spring steel ⅛ to 3/16″ wide and slips through the keyway for direct manipulation of the bolt.

Open keyways are generally used, as opposed to blind, or masked, keyways. The keyway runs from the face through the plug body. You can open the lock with relative ease. Use the tool shown in Fig. 15-7 to work the bolt directly.

During the past decade, new standards have been set for filing-cabinet locks; many manufacturers have revised their dies to do away with the open keyway and the breach in security it represents. This means that today many filing-cabinet locks have bolts that are isolated from the keyway. A few manufacturers use their original dies but add a fifth pin. This pin is longer than the others and has no upper chamber. It cannot be lifted and so blocks access to the bolt. Another option has been to block the keyway end with a horizontal pin. The jimmy no longer works, and we must turn to the more traditional methods of opening a keyless lock. There are three choices: pick the lock, drill the plug, or force the bolt.

Picking is the most desirable of these alternatives, but requires patience and skill. Drilling is fast and means the loss of the lock, but it is preferable to forcing the bolt. If the customer accepts the risk, you can pry one drawer open far enough to manipulate the bolt with a letter opener.

Chapter 16

Automotive Locks

While the automobile has been with us since the start of the century, the lock was slow to be adopted. However, by the late 1920s nearly every auto had an ignition lock, and closed cars had door locks as well. Current models may be secured with half a dozen locks.

SIDEBAR LOCKS

While automobiles may be fitted with pin- or disc-tumbler cylinders, the sidebar lock is most typical and the only one that merits special attention. The other two have been discussed earlier.

Figure 16-1 illustrates this lock in simplified form. Notice that the wafers have V-shaped notches in their sides. Inserting the correct keys aligns the notches, and the spring-loaded sidebar moves out of the cylinder and into the plug. Once the sidebar passes the shear line, the plug is free to rotate.

KEY FITTING

Key fitting for disc- and pin-tumbler locks has been described in earlier chapters. The techniques used for automotive variations of these locks are no different.

Key fitting sidebar locks requires that the lock be disassembled. Remove the lock from the vehicle and, with a small screwdriver or other sharp tool, release the staking that holds the sidebar and associated spring in place.

Remove the tumblers and their springs. Insert the appropriate blank into the keyway and replace the tumbler at the far end of the plug. Working from the sidebar chamber, scribe a mark on the blank where the blank touches the tumbler. This is the guide mark for the cut. File the mark lightly and reinsert the blank; the file mark should be at or under the tumbler. File slowly until the V-notch on the side of the tumbler appears to align with the sidebar. Hold the sidebar and spring temporarily in place with your fingers and try the key. If all is right, the sidebar will extend into the notch, freeing the plug. The key should turn.

Repeat this operation for each of the remaining tumblers.

DOOR LOCKS

Modern door locks are disassembled from inside the door. Figure 16-2 shows a typical example. Remove the arm rest (shown as No. 7 in the drawing) and inside handles. Most handles are secured by a spring clip that is accessible with the tools shown in Fig. 16-3. Note the

Fig. 16-1. A simplified drawing of a sidebar lock.

positions of the handles on their spindles for assembly.

Remove the Phillips screws holding the door trim (6) to the door body (2). Gently pry the trim away from the body and set it aside. The door lock mechanism is visible through access holes in the inner door-body panel.

Most problems involve the lock mechanism and not the cylinder. Check for broken springs, missing retainer clips, and rust on working parts. The cylinder (20) is secured by a U-shaped retainer clip. Cylinders can be disassembled, although the modern practice is to replace the cylinder as a unit, rather than to attempt repairs.

Key to Fig. 21-2

(1) Door hinge (upper)
(2) Door body
(3) Damper
(4) Door hinge (lower)
(5) Door weatherstrip
(6) Door trim
(7) Arm rest
(8) Door glass
(9) Main roller guide (front)
(10) Channel
(11) Regulator assembly
(12) Regulator handle
(13) Upper stopper (lower)
(14) Sub-roller guide
(15) Upper stopper
(16) Main roller guide (rear)
(17) Door inside handle
(18) Door inside lock
(19) Door outside handle
(20) Cylinder lock
(21) Door lock
(22) Door striker
(23) Shim

Fig. 16-2. A typical front-door assembly. (courtesy Chrysler Corp.)

Fig. 16-3. These K-D door handle tools are available from auto parts stores. View A illustrates the tool for Chrysler retainers; view B, the tool for Ford and General Motors.

The striker can be moved to align the door with the body panels and to compensate for door sag. Loosen the mounting screws a quarter turn or less and, using a soft mallet, tap the striker into position (Fig. 16-4). Shims are available from the dealer to move the striker laterally, out from the door post.

While this is out of the province of locksmithing, as such, it is useful to know that the door can be adjusted at the hinge on all automobiles except Vega (Fig. 16-5). Loosen the pillar-side bolts slightly and manipulate the door.

IGNITION LOCKS

Early production ignition locks are secured to the dash by bezel nuts or retainer clips and feature a "poke hole" on the lock face or in the back of the cylinder that

Fig. 16-4. Door strikes are adjustable. (courtesy Chrysler Corp.)

Fig. 16-5. The door adjustment is at the pillar side of the hinges for most vehicles. (courtesy Chrysler Corp.)

Fig. 16-6. The glove-box cylinder can be removed from the outside with the tool shown. (courtesy Ford Motor Co.)

gives access to the retaining ring. In a few cases, the lock is held by screws from the underside. Late-model locks, (the kind that lock the steering wheel as well as the ignition) are not something that the average locksmith should become involved with. These locks can be serviced, but special tools and special knowledge are required. The upper portion of the steering column must be dismantled and, in some cases, the steering column must be dropped from the dash hanger. A mistake in assembly—even so much as using the wrong screw—can defeat the collapsible steering column. If you want to get into this aspect of the trade, by all means spend a few weeks at dealer schools and invest in the proper tools.

GLOVE COMPARTMENT LOCKS

Modern glove-compartment locks are typically se-

cured by a retainer accessible from the front of the lock. Figure 16-6 shows how this is done. Make a ¼″ hook in the end of a piece of Bowden (hood-latch) cable. Insert the hook into the keyway and work the retainer free. Most are installed with the open ends toward the passenger door; nudge the retainer toward the center of the car to disengage it.

Extract the cylinder with the key and release the bolt by hand. The cylinder is stamped with a numerical keying code—the same code that is used for the trunk or, in the case of station wagons, the tailgate.

TRUNK LOCKS

Figure 16-7 shows a luggage-compartment latch, typical of those found on middle-priced automobiles. More expensive cars use a more complex lock mech-

Fig. 16-7. The trunk lock assembly used on Mustang and Cougar autos. (courtesy Ford Motor Co.)

anism that may be solenoid-tripped from inside the vehicle. The basic parts and their relationships are the same. The cylinder is secured to the trunk lid by means of a retaining clip and works the lock assembly through a tailpiece or extension. (A few cars mount the cylinder on the lower body panel and the striker on the lid.) The striker is located on the aft body panel.

Most lock assemblies are adjusted only at the striker plate. The plate is mounted on elongated bolt holes and can be moved up, down, left, and right. Other types have an additional adjustment at the latch. The idea is to position the parts so the lid closes without interference from the striker and tightly enough to make a waterproof seal between its lower edge and weatherstripping.

Chapter 17

Decoding Locks, Lockpicking, and Emergency Entry

Another skill required by the locksmith is the ability to cut different key types when only the lock manufacturer and key number are known. The locksmith may never see the lock or key; at best he may have only the lock. Certain information may be required of the customer, depending on city or county ordinances. In most cases you will need to find out who owns the lock. Has the owner given authorization to have a key made for the lock? Requests involving high-security locks must be forwarded to the factory together with a written release from the owner.

The code varies with each manufacturer, but includes this information:

☐ Key blank type.
☐ Cut spacing.
☐ Cut depths.

READING CODES

Direct-digit and indirect codes are most popular. The direct-digit code consists of one or two letters followed by a series of numbers. The letters identify the lock series, and the numbers refer to the depths of the cuts. The series may start at the last tumbler (the one closest to the key tip) or at the first (next to the key shoulder). Indirect codes are safer: the numbers must be decoded before the cuts can be identified. Professional locksmiths have the necessary code books.

The smallest number—either 0 or 1—in both kinds of codes represents the shallowest cut; 9, the last of the series, represents the deepest cut. Once you have the number, you can usually determine the type of lock. For example, key code LL76 is limited to desks. This fact can be useful in that it narrows the range of possible blanks.

The code book says that LL76 is a Yale desk lock and that you can use ILCO 01122A, National Y12, or Yale 9278A blanks. The depth and spacing charts show five depths, each 0.020″ apart, and five spacings. Counting from the tip to the shoulder, the first four cuts are spaced 0.094″ apart; the last cut is 0.140″ from the shoulder. Code number LL76 translates as 24253 and is read from the tip to the shoulder. To cut a key with this code, place a Yale No. 2 depth and spacing key in the vise, together with the appropriate blank. Cut the No. 2 bits. Repeat the operation with Nos. 3, 4, and 5 depth and spacing keys.

Depth and Spacing Keys

Depth and spacing keys simplify the work. These keys are milled to various depths and spacings. Because manufacturers vary these dimensions, you need a complete key set, or at least one that covers the major American locks.

Schlage Wafer-Tumbler Keys

Schlage keys are easy to cut by code since the cuts all have the same depths. Three Schlage keys and one depth key is all that is required.

LOCKPICKING

Lockpicking is the ultimate test of a locksmith; it tries his knowledge as well as his skill. To many customers, lockpicking is the one ability that distinguishes a true locksmith from someone who merely tinkers with locks.

LOCKPICKING TOOLS

Basic lockpicks are illustrated in Fig. 17-1 and include the following:

☐ Half-round feeler
☐ Round feeler
☐ Rake
☐ Half-rake
☐ Diamond
☐ Double-diamond
☐ Circular
☐ Reader
☐ Extractor
☐ Mailbox
☐ Flat lever

You can make your own lockpicks from flat cold-rolled steel, 0.020 to 0.025″ thick. The strip should be about 6″ long and at least ⅓″ wide. One end may be fitted with a handle, or both ends may carry working surfaces.

Picks for warded locks come in various designs also. A few of them are shown in Fig. 17-2. It's usually a simple matter to make one of these picks. Sometimes a standard precut key shaved to pass the keyhole wards is all that is needed.

Tension tools for use with various picks are made of hardened spring steel. Normally they are from three to five inches in length, excluding the parts that actually fit into the keyway opening.

A letter box tension tool is made from a thin strip of spring steel about four inches long and has a ⅛″ to ¼″ probe at its working end. It is best to have several with different probe lengths to ensure that you always have the correct one on hand.

Picks and tension tools for double-sided disc-tumbler locks are unique in their design (Fig. 17-3). They

should be purchased from your locksmith supply house; it's cheaper to buy them than to make them.

Lever lockpicks come in two varieties—those for true lever locks and those that double as key picks. The true lever lockpick is made from 0.095″ or thinner spring steel and has a tension tool that is the same thickness (Fig. 17-4). The second variety is cut from spring steel, or made from standard key blanks.

Note: It is illegal to possess lockpicks without a locksmith's license in many jurisdictions. The penalties are severe, even for first offenders.

Fig. 17-1. Basic lockpicks.

Fig. 17-2. Picks for warded locks.

LOCKPICKING TECHNIQUES

The prerequisite for successfully picking a lock is a sound knowledge of how the lock mechanism works. Without such an understanding, picking a lock becomes a matter of sheer luck.

Picking the Pin-Tumbler Lock

To pick a pin-tumbler lock you must be able to determine when the cylinder pins are at the shear line. You must apply just the right amount of pressure on the cylinder core with a tension tool. Figure 17-5 shows a feeler pick raising a pin to the shear line. To raise the pins in this way requires constant practice. Too much or too little pressure on the pick will cause the pin to stick at the wrong place (Fig. 17-6).

Factory precision in drilling the cylinder pin holes is never perfect. An enlarged view of the top of a plug would show that the holes are not really in alignment (Fig. 17-7). This is of help in opening the lock. Since the holes are not perfectly aligned, you can hold one pin at the shear line while working on others. Another built-in advantage is the fact that locks always have at least some clearance between moving parts—and a little working space is all a good locksmith needs to pick a lock.

In raising the various pins, it is preferable to start with those of the greatest length. By using this method, you progress from the smallest amount of pick movement

Fig. 17-3. Picks for double-sided disc-tumbler locks.

0.095 in.
SPRING
STEEL
(FLATTENED
AT
POINT)

A

B

Fig. 17-4. A tension tool (A) and picks for warded locks (B).

Fig. 17-5. A feeler pick raising a pin to the shear line. (courtesy Desert Publications)

up to the greatest. As each pin reaches the shear line, the core moves ever so slightly; you will be able to feel this movement when it is transferred to your tension tool.

Using a feeler pick, move one pin at a time. Since the purpose of the pick is to move the individual pins up *to* the shear line, not *above* them, you must take your time. If you feel that all but one of the pins have reached the shear line, and the last one just won't go, you have probably pushed the last pin above the shear line. Release the tension and start again, with slightly less tension than before.

Remember that you must vary the amount of pressure you apply to the pick, depending upon the amount of resistance you receive from each pin. Also, different locks require different amounts of pressure. What works on one lock will not necessarily work on the next. Even two locks of the same make and style may not require the same amount of pressure.

It is sometimes possible to use picks (especially the rakes) to "bounce" the pins to the shear line. To do this, insert the pick fully then withdraw it quickly while holding light tension on the plug. Simply forcing the pick rapidly in and out of the cylinder won't do the job: Such a technique merely bounces the pins *above* the shear line.

Picking Warded Locks

Warded locks are very easy to pick. Sometimes you can even use a pair of wires: one for throwing the bolt, the other for adjusting the lock mechanism to the proper height for the bolt to be moved. Here, as with any other kind of lockpicking, it is a matter of practice on a variety of locks.

Warded padlocks have only one or two obstacles, thus simplifying your choice of pick. Some warded locks (such as the Master) which have keyways with corrugated cross sections can be opened in no time at all by having several precut blanks available. The regular picks are usually just a fraction of an inch too wide to fit into the keyway.

Smaller locks with correspondingly smaller keyways require that you either get a second set of tools and cut them down to size, or use several blanks to obtain the proper picks for these locks.

Picking Lever-Tumbler Locks

Figure 17-8 shows a lock with the frontplate removed, indicating the position of the tension tool in relation to the bolt. Holding the levers with the tension tool (as shown in the side view) enables you to manipulate the levers with the pick.

Fig. 17-6. Too much pressure with a pick will raise a pin above the shear line.

Fig. 17-7. Cylinder pin holes are never in perfect alignment.

In learning to pick the lever lock, it is best to start with a lock with only one lever in place. At first, work on the lock with the faceplate removed so you can get an idea of how much pressure to apply to the bolt, and how much movement is required of the pick to move the lever into position for the bolt to move through the lever gating.

As you progress to locks with several levers, keep the faceplate on. When you encounter a problem, remove the faceplate so you can see what you're doing wrong. Always remember to insert the tension tool first. Push it to the lowest point within the keyway so the pick will have maximum working space.

As with the pin-tumbler lock, one tumbler tends to take up most of the tension; work on this one first. As you move this tumbler slowly upward to the proper position for the bolt to pass through the gate, you will feel a slight slackening in tension from the bolt as it attempts to force its way into the gating. This tension will be transmitted to you through the tension tool. Stop at this point and do the same to the next lever with the greatest amount of tension. Continue until all the levers reach this point. Then by shifting the tension tool against the bolt, the bolt will pass through the gate, opening the lock.

There must be at least some pressure on the tension tool at all times. A lack of pressure will cause any levers in position to drop back into their original locations.

Picking Disc-Tumbler Locks

Standard disc-tumbler locks (those using a single-bitted key) take the same picks as pin-tumbler locks. Also like the pin-tumbler lock, a disc-tumbler lock can be picked by bouncing the tumblers to the shear line. Usually a rake is the best tool for this job, but other picks can be used too.

In order to develop your proficiency, you should also try opening the disc-tumbler lock with the feeler pick, working on each tumbler individually. By doing this you learn which picks you are most comfortable with and

which ones you prefer for different types of locks.

When working on double-bitted disc-tumbler locks, the bounce method is best. Insert the tension tool into the keyway and apply a slight pressure to the core as the pick is pulled out of the lock. These locks can also be opened with a standard pick set and tension tools, but it can take from ten minutes to half an hour.

LOCKPICKING CONSIDERATIONS

The following are points to remember when picking any lock:

☐ It is to your advantage to use the narrowest pick available in order to give yourself maximum working space.
☐ Hold the pick as you would a pen or pencil. Don't use wrist action; your fingers work much better in manipulating the pick in the lock.
☐ Steady your hand with your little finger against the door. When working on a key-in-knob cylinder, steady your hand against the edge of the knob.
☐ Your pick should enter the keyway above the tension tool without moving any of the tumblers. If it doesn't, the tension tool is either too high in the lock or the keyway grooves are such that the tension tool must go at the very top of the keyway.
☐ Worn cylinders and loose plugs frequently open easier and quicker with the bounce method—normally within three to five tries.

EMERGENCY ENTRY PROCEDURES

Sometimes locksmiths gain entry into locked premises by picking open the lock. But a good locksmith can also gain entry through special techniques called Emergency Entry Procedures (EEP). In some instances, a lock resists picking. It is then that the EEP are used.

Fig. 17-8. A tension tool exerting pressure on lever tumblers.

Fig. 17-9. A cylinder removal clamp. (courtesy Desert Publications)

Emergency Entry Procedures are also called *special techniques, emergency measures*, and *quick entry*. Much of the information and illustrations in this chapter come from Desert Publications' book *Lock Out*.

As a professional locksmith, you should have the proper equipment to make emergency entries as quickly and as neatly as possible. Some entry tools are manufactured; you should purchase a set of these for your use. Some tools you can fashion yourself. Experience will show you which tools and techniques work best for you.

Using Emergency Entry Procedures can account for a substantial percentage of a locksmith's income, thus it is wise to be knowledgeable in a variety of entry techniques. It may not seem like good locksmithing practice to force a lock open, but economics, time, and other requirements may make it necessary.

Several preliminaries should be undertaken when preparing to make an entry using EEP. Try to obtain as much information as possible. With this information, you can decide what methods will work. Ask questions about the type of lock the key numbers, nearby open windows, a possible extra key held by someone else, the condition of the lock, alarms and other devices attached to the door, etc. All this information will aid you in preparing for the entry. Fortunately, most lockouts require only that you pick the lock or make a new key.

DRILLING PIN-TUMBLER LOCKS

Drill a lock only as a last resort. There are two drilling methods used with pin-tumbler locks. The first method is the drilling of the cylinder plug. This has the advantage of saving the cylinder itself. The inner core can be replaced. This method destroys the lower set of pins just below the shear line, allowing the plug to be turned. Follow these procedures for this method:

1. Drill the plug below the shear line.
2. Insert a key blank and a wire through the drill hole to keep the upper pins above the shear line and the destroyed pins below it.
3. Turn the cylinder core and open the lock.
4. Once the door is open, dismantle the lock; remove and replace the core and fit new pins to the lock. If you used the plug follower when removing the core, you will only have to fit new lower pins to the lock to match a key.

The second method involves drilling just above the shear line into the upper pins.

1. Insert a key blank into the lock first to force all the pins to the upper parts of their chambers.
2. Drill about ⅛″ above the shear line or shoulder of the plug and directly above the top of the keyway (Fig. 17-9). You can use a drilling jig to ensure drilling at the proper point on the lockface. Use a ⅛″ or 3/32″ drill.
3. Poke a thin wire or needle into the hole and withdraw the key to just inside the keyway. You can then use the tip of the key to turn the core. Removing the key partway allows the bottom pins to drop below the shear

Fig. 17-10. A cylinder removal clamp. (courtesy Desert Publications)

line, while the wire or needle keeps the upper pins above the shear line.

4. Once the door is open, dismantle the lock, remove the cylinder, and replace it with a new one.

CYLINDER REMOVAL

Sometimes the cylinder must be removed in order to open a lock. This means shearing off the screws that hold the cylinder in place. Several different rim cylinder removal tools are available. The one in Fig. 17-10 can be made in your own shop. It's called a cylinder removal clamp.

To make one, follow these steps:

1. Use a piece of steel tubing about ⅛" larger in diameter than the cylinder. The tubing should be no more than 2½" long.

2. Cut into the tubing as shown in Fig. 17-10 about 1 3/16".

3. Cut a center hole at the end of both cuts ¼" to ⅓" in diameter and insert a steel rod about 10" long. The rod becomes your handle for turning the cylinder.

4. Drill two holes in the tubing so you can insert a bolt perpendicular to the handle. Ensure that the bolt will be ½" in front of the handle. The bolt threads must extend on both sides of the tubing.

To remove the cylinder, follow this procedure:

1. Set the removal clamp over the edge of the cylinder after first removing the cylinder rim collar. Tighten down the tension bolt. This provides the gripping pressure to the cylinder.

2. Twist the cylinder clamp by the handles and

force the cylinder to rotate, shearing off the retaining screw ends.

3. Remove the cylinder from the lock. When removed, reach in and open the door by reversing the bolt.

You can also use the standard Stillson wrench to remove the cylinder.

Cylinders can also be pulled out of the lock unit, but this method ruins the cylinder threads, requiring you to replace the entire unit. The tool used is called a nutcracker. The sharp pincer points are pushed in behind the front of the cylinder face. Clamp down and pull the cylinder out. Many times an unskilled locksmith can ruin the entire lock and a portion of the door by not knowing how to use this tool properly.

Some cylinders have to be drilled. Follow this procedure:

1. Drill two 3/16" in. diameter holes about ⅞" to 1⅛" apart on the face of the cylinder.

2. Drive two heavy bolts into these holes so that at least 1½" is sticking out.

3. Place a pry bar or heavy screwdriver between the bolts and use it as you would a wrench to force the cylinder to turn, shearing off the long screws.

WINDOW ENTRANCES

Window entrances are relatively easy. The old butter knife trick is usually successful. Since most window latches are located between the upper and lower windows, sliding a knife up between the windows allows you to open the latch.

Should the area be too narrow for a knife, shim, or other device, drill a 1/16" hole at an angle through the wood molding to the base of the catch. Insert a stiff wire and push back the latch.

OFFICE LOCKS

Most office equipment can be opened using a few basic methods and a handful of tools.

Filing Cabinets

Though most filing cabinets have the lock in essentially the same position, the locking bar arrangement for the various drawers will vary. These variations have to be considered in working on the locks.

One method is to work directly on the lock itself:

1. Slide a thin strip of spring steel ⅛" thick into the keyway and pull the bolt downward. This will unlock the lock (Fig. 17-11).

2. If the lock has a piece of metal or pin blocking access to the locking bolt, use a piece of stiff wire with one end turned 90°. Insert the wire between the drawer and the cabinet face and force the bolt down with the wire. This will allow you to open the drawers.

3. If there is not enough room for you to work with the wire, use a piece of thin steel or a small screwdriver to pry back the drawer from the cabinet face to allow you to see and work with the wire.

You can also open the drawers individually if necessary:

1. Use a thin piece of spring steel or a wedge to spread the drawer slightly away from the edge.

2. As your opening tool use another strip of steel about 18″ long, ½″ to 1″ wide, and 0.020″ thick.

3. Insert the opening tool between the drawer catch and the bolt mechanism (Fig. 17-12).

4. With a healthy yank, pull the drawer open. The opening tool creates a bridge for the drawer catch to ride upon and pass the bolt.

Other filing cabinets can be inverted to release gravity-type vertical engaging bolts. When the lock mechanism itself is fouled up, the best way to proceed would be to drill out the cylinder and replace the entire assembly.

Desks

Desks with locking drawers controlled from the center drawer can be opened in a couple of ways besides picking and drilling.

Look at the desk from underneath to see what the locking mechanisms for the various drawers looks like. Notice that the locking bar engages the desk by an upward or downward pressure, depending upon the bolt style. The bolt is pushed into the locked position by the motion made when closing the desk drawer all the way. Herein

Fig. 17-11. Sometimes a thin strip of spring steel inserted into the keyway of a filing cabinet lock (A) can be used to pull the bolt down (B). (courtesy Desert Publications)

lies the weakness in the desk's security. The bolt usually needs to be pushed up from under the desk by hand to open most of the drawers. The center one has its own lock.

In other desks you may have to use a little force and pull outward on the center drawer to push the bolting mechanism downward slightly to open the various drawers.

To open the center lock, follow this procedure:

1. Use two screwdrivers and some tape or cardboard. Put the cardboard or tape between the drawer and the underside of the desk top so you don't mar the desk.

2. Insert a screwdriver and pry the drawer away from the desk top.

3. With the other screwdriver pull the drawer outward to open it. With practice, this can be done with only one screwdriver.

If you drill a small hole in the drawer near the lock, you can insert a stiff piece of wire, such as a paper clip, to push down the plug retainer ring. In doing so, you pull the plug free of the lock, causing the bolt to drop down into the open position.

DOORS

Possibly the simplest way to open most doors is with a pry bar and a linoleum knife:

Fig. 17-13. Opening a door using a pry bar and linoleum knife. (courtesy Desert Publications)

Fig. 17-12. Sometimes the drawer of a filing cabinet (A) can be opened by wedging the drawer away from the cabinet frame (B) and inserting a strip of steel between the drawer catch and the bolt mechanism (C). (courtesy Desert Publications)

Fig. 17-14. Opening a door with a Z-wire. (courtesy Desert Publications)

3. When this is done, bring the linoleum knife forward, pushing the latchbolt into the locking assembly, and opening the door. If there is no safety catch, the knife alone can be used to move the bolt inward. It's also possible to use a standard shove knife or even a kitchen knife.

Sometimes the deadlatch plunger is in the lock but there isn't room to insert a pry bar. What do you do? Use wooden wedges. Insert one of them on each side of the bolt, about 4 to 6 inches from the bolt assembly. Spread the door away from the jamb. Then use a linoleum knife to work the bolt back.

Some doors and frames have such close clearances that you cannot insert a wooden wedge or pry bar. In an instance such as this you can use a stainless steel door shim. Merely force it into the very narrow crevice between the door and the frame and work back the bolt.

Often a door lock can be opened with a Z-wire. This tool is made from a wire at least 0.062″ thick and ten to twelve inches long (Fig. 17-14). The Z-wire is inserted between the door and the jamb. When the short end is all the way in, it is rotated toward you at the top. As you do this, the opposite end will rotate between the door and the jamb. It will contact the bolt and retract it. If the bolt should bind, exert pressure on the knob to force the door in the direction required.

Sometimes you may be required to open locked

1. Insert the linoleum knife between the door and the jamb with the point tipped upward.

2. Insert the pry bar as shown in Fig. 17-13. Exerting a downward motion on the pry bar spreads the door slightly and allows the locking safety latch to be disengaged.

Fig. 17-15. The rubber band technique is usually the easiest way to unlock most chain locks.

Fig. 17-16. A variation on the rubber band technique.

Fig. 17-17. Bent coathangers used to lift automobile pushbutton door locks.

chain latches. You can, of course, force the door and break the chain, but there are better ways. The rubber band technique works most of the time:

1. Reach inside and stick a tack in the door behind the chain assembly (Fig. 17-15).

2. Attach one end of a rubber band to the tack; attach the other end to the end of the chain.

3. Close the door. The rubber band will pull the chain back. If it doesn't pull the chain off the slide, shake the door a little.

4 If the door's surface will not receive a tack, use a bent coathanger to stretch the rubber band (Fig. 17-16). Make sure the coathanger is long enough and bent properly so you can close the door as far as possible.

If the door and jamb are even and there is enough space, a thin wire can be inserted to move the chain back.

Sometimes you can open a door very easily if it has a transom. You can use two long pieces of string and a strip of rubber inner tubing. The tubing should be eight to ten inches long. Attach a string to each end of the tubing so you can manipulate the tubing from the open transom. Lower the tubing and wrap it around the knob. Pull up firmly on both strings to maintain tension and turn the knob. This method can be used with either a regular door knob or an auxiliary latch unit.

AUTOMOBILES

Automobile EEP can be used on front and rear ventilation windows, door windows, doors, and trunks. Automobile locks vary considerably in their design and reliability. Despite this, you can use several methods to gain access. Most of them are faster and easier than picking the lock. In some automobile locks, picking must be ruled out entirely because of the sidebar cylinder which makes picking a very time-consuming, tedious job.

Keyless entry into an automobile is most commonly done at the windows, the most vulnerable part of a car's security. The door release pushbutton lever is often the easiest area to attack, especially on the newer car models that have a single pane of glass. The procedure is fairly simple:

1. Use a straightened coathanger with a loop or triangle in its end as shown in Fig. 17-17.

Fig. 17-18. Sometimes a paint scraper can be used to make room for a coathanger that can pull open a car pushbutton lock. (courtesy Desert Publications)

248

Fig. 17-19. A .22 caliber gun cleaning set. (courtesy Desert Publications)

2. If the window is rolled up tight, force a paint scraper between the edge of the window and the weatherstripping (Fig. 17-18).

3. Insert the coathanger in the crevice made by the paint scraper and pull the pushbutton lever up with the loop or triangle. Of course, a coathanger isn't the only tool that could work here. Any instrument that could slip through the crevice and lift the lever could do the job.

Two .22 caliber gun cleaning sets can help you lift either the pushbutton lever or the door handle (Fig. 17-19). The cleaning tools come in sections so you can make a variety of rod lengths.

1. Use the two sets to make a rod long enough to reach across the inside of the car.

2. Use the slotted cleaning attachment and some fishing line to make a loop in the end of the rod (Fig. 17-20).

3. Work on the window or door opposite the side you are on. Raise the lever or the handle with the loop by pulling on the line and lifting the rod up.

The rear ventilation, or wing, window comes in two varieties: with and without a locking button on the swivel.

Without the locking button, gaining entry is only a matter of forcing the swivel lock up into the unlocked position.

To open rear ventilation windows without locking buttons on the swivel, follow this procedure:

1. Insert a putty knife or paint scraper between the window and the frame, bending the knife slightly to allow the unlocking tool to enter. In this case, the unlocking tool could be a thin piece of wire.

2. Loop the wire around the swivel latch and pull upward. The wire should be bent slightly on the end to ensure a better upward motion without it slipping off the lever.

Most cars now have locking pushbuttons in addition to the swivel levers. To unlock these mechanisms you need standard automobile opening tools. These tools are inserted on different sides of the lock (Fig. 17-21). One is used to depress the button by pulling the tool towards you. The other is twisted slightly to push the lever into the unlocked position.

The front ventilation window gives you access to the window roller handle the door handle. Two different tools

Fig. 17-20. A long rod with a closable loop on the end makes a good car lock opening tool. (courtesy Desert Publications)

Fig. 17-21. Standard automotive opening tools are used to open rear ventilation windows with locking buttons. (courtesy Desert Publications)

Fig. 17-22. Tools for opening the front ventilation windows. (courtesy Desert Publications)

In some trunk cylinders (there is a retainer. An L-shaped wire inserted into the keyway will catch the retainer, and you can pull it down and pull the plug out of the lock (Fig. 17-24). After that a small screwdriver or an awl can be used to push back the catch mechanism.

The easiest of all automobile locks to open is the glove compartment lock. Insert an L-shaped wire into the keyway and pull down the retainer. This wire actually works on two distinct types of locks. In one, you remove the plug; in the other, the wire engages the glove compartment catch mechanism and opens it with a downward movement.

FORCED ENTRY

Forced entry in place of proper professional techniques is never recommended, except in emergencies or when authorization is given by the owner. There are some simple rules about forced entry you should consider:

☐ Attempt forced entry only as a last resort. Try other techniques first.
☐ If you must jimmy the door, do it carefully. Today antique doors and frames can be only part of a valued collection that may be in a home. Don't risk a costly lesson by being too hasty.

can be used, depending upon the amount of space you have to work with (Fig. 17-22). Both tools work something like the one shown in Fig. 17-23. The window must be pried slightly to insert the tool.

A Lenco car opener tool can also be used from the front ventilation window. You can get the tool through your local locksmithing supply house. The tool can pay for itself with the first lockout that you are called upon to open.

How you gain entry into the trunk of a car depends upon where the lock is located. Drilling is an acceptable practice here. Remember that door locks and the ignition usually take the same key; the glove compartment and the trunk usually take the same key also. Sometimes you can open the trunk lock by inserting the door key and pulling it out as you turn it.

Picking the trunk lock is simple, but it can be time consuming if you have not worked with very many automobile locks. You can knock the lock out by forcing it inward with a hammer and a heavy screwdriver, but this ruins the lock and, sometimes, associated parts on the inside.

Drilling can also have its drawbacks. You can drill out the plug by shearing off the pins. This means replacement of the lock unit. Lately, manufacturers have improved the security of trunk locks by installing a steel plate in front of the catch to prevent drilling. However, it is possible to drill about 2″ to the side of the catch bolt. This enables you to use a bent piece of wire to force the catch backwards to open the trunk.

Fig. 17-23. One method of opening the front ventilation window. (courtesy Desert Publications)

250

☐ If you must break a window to gain access to a lock, break a small one. Replacing broken windows can be expensive.
☐ Don't saw the locking bolt. This is very unprofessional.

Remember that lockpicks and many other entry tools are considered burglar tools in some jurisdictions. In many cases you must have a license to carry them. Check with your local police, keep your locksmith's license current, and when a tool is worn out, destroy it.

Be aware of the trust which your city and customers

Fig. 17-24. An L-shaped wire used to open trunk locks.

put in you by allowing you to be a locksmith and possess certain tools and knowledge that others do not have. Don't abuse this privilege.

Chapter 18

Combination Locks

Like other locks, the combination lock is available in a variety of types and styles. In this chapter we will consider a number of representative types. Some of these locks are more secure than others, but each has its role. For example, it would be inappropriate to purchase a high-quality combination padlock for a woodshed; by the same token, it would be foolish to protect family heirlooms with a cheap lock. Even the best lock is only as good as the total security of the system. A thief always looks for the easiest entry point.

PARTS

Though combination locks have some internal differences, all operate on the same principle—rotation of the combination dial rotates the internal wheel pack. The pack consists of three (sometimes four) wheels. Each wheel is "programmed" to align its gate with the bolt-release mechanism after so many degrees of rotation. Programming may be determined at the factory, or it may be subject to change in the field. As the wheels are rotated in order (normally three turns, then reverse direction for two turns, and then reverse direction again for one turn), the gates are aligned by stops, one for each wheel and one on the wheel-pack mounting plate. When all the gates are aligned, the bolt is free to release. Padlocks are built so the shackle disengages automatically, or manually by pulling down on the lock body. The captive side of the shackle, the part known as the *heel*, has a stop so the shackle will not come free of the lock body.

One weakness of low-priced locks is their ease of manipulation. With practice you can discriminate between the clicks as the wheels rotate. Once you learn to do this, manipulation of the lock is child's play. Fortunately, manufacturers of better locks make false gates in the wheels, and these gates make manipulation much more difficult. True, a real expert will not be thwarted, but experts in this arcane art are few and far between. It takes training and intensive practice to distinguish between three or more false gates on each wheel and the true gate.

MANIPULATION

Manipulation of combination locks is a high skill—almost an art. A deep understanding of combination-lock mechanisms and many, many hours of practice are needed to master this skill. But the rewards—both in terms of the business this skill will bring to your shop and the personal satisfaction gained—are great.

No book can teach you to manipulate a lock any more than a book can teach you to swim, box, or do anything else that is essentially an exercise in manual dexterity. But a book can teach the rudiments—the skeleton, as it were, that you can flesh out by practice and personal instruction from an expert.

At the risk of redundancy, allow me to repeat that manipulation is a matter of touch and hearing. It also involves an understanding of how the lock works. Electronic amplification devices are available to assist the locksmith. These devices are useful, especially as a training aid.

Padlocks usually do not have false gates and so are ideal for beginners (Fig. 18-1). Pull out the shackle as you rotate the wheels. This tends to give better definition to the clicks. Work only for one number at a time. Stop when the bolt hesitates and touches the edge of the gate. The bolt has touched the far side of the gate, so move back a number on the dial and note it. Turn the dial in the reverse direction, going past the first number at least twice. Pull

on the shackle and slowly continue to turn the dial. When you sense that the bolt has touched the gate, note the number that came up just before the bolt responded. This is the second number of the combination. The third number comes easier than either the first or second one. The dial may stop at either the first or second number. Since many of the earlier combination locks did not have the accuracy of modern locks, the bolt catches at any one of the three numbers at any time.

You may have the three numbers, but not in their correct order. Vary the sequence of the numbers until you hit the right one.

Because of the imperfections in older and inexpensive modern locks, the bolt may stop at points other than the gates. Only through practice can you learn to distinguish between true and phantom gates.

For practice, obtain two or three locks of the same model. Disassemble one to observe the wheel, gate, and bolt relationship and response. These insights will help you manipulate the other two locks.

Fig. 18-1. A typical combination padlock in exploded view. (courtesy Desert Publications)

Fig. 18-2. The Simplex pushbutton combination lock. (courtesy Simplex Security Systems, Inc.)

DRILLING

All combination locks can be opened by drilling as a last resort. To drill these locks follow this procedure:

1. Drill two ⅛″ holes in the back of the lock.

2. Turn the dial and determine that the gate of one wheel is aligned with one of the holes. You may be able to see the gate. If that fails, locate the gate with a piece of piano wire inserted through one of the holes.

3. When the gate and the hole are aligned, note the number on the dial face. Determine the distance, as expressed in divisions on the dial, between the hole and bolt.

4. Subtract this distance from the reading when the gate is aligned with the hole. The result is the combination number for that wheel.

5. Reverse dial rotation and find the number for the second wheel. Do the same for the third.

CHANGING COMBINATIONS

Many locks are designed for combination changes in the field. I will discuss three of them.

Fig. 18-3. Changing Simplex combination, Step 2.

Fig. 18-4. Changing Simplex combination, Step 3.

Fig. 18-5. Changing Simplex combination, Step 4.

Fig. 18-6. Changing Simplex combination, Step 5.

Fig. 18-7. Changing Simplex combination, Step 6.

Sargent and Greenleaf

S & G padlock combinations are changed by key.

1. Turn the numbers of the original combination to the change-key mark located 10 digits to the left of the zero mark on the dial face.

2. Raise the knob on the back of the lock to reveal the keyway. Insert the key and turn 90°.

3. Repeat step 1, but use the new combination and the change-key mark as zero.

4. Remove the change key and test the new combination.

Simplex

The Simplex is a unique combination lock, employing a vertical row of pushbuttons rather than the more usual dial (Fig. 18-2).

1. Turn the control knob left to activate the buttons.

2. Release the knob and push the existing combination (Fig. 18-3).

3. Push down the combination change slide on the back of the lock (Fig. 18-4).

4. Turn the control knob left to clear the existing combination (Fig. 18-5).

5. Push the button for the new combination—firmly and in sequence (Fig. 18-6).

6. Turn the control arm right to set the new sequence (Fig. 18-7).

Dialoc

Partial disassembly is required to change the combination. Refer the parts numbers in the following steps to Fig. 18-8.

To disassemble:

1. Remove the Dialoc from the door by withdrawing the two mounting screws 487-1, holding the inside plate 821 against the door.

2. Place the outside plate 820 face down on a smooth work surface so that housing 414 is facing up.

3. Remove the three nuts 484-5 holding the housing cover 745.

4. After removing the nuts, gently lift the housing cover up and away from the housing. Take care that control-bar spring, 744 is not dislodged and lost in the process. Lay it where it won't get lost.

5. Now, remove the secondary arm 402.

6. Grasp the end of the control bar 726 and lift out the four ratchet assemblies. Observe how they are arranged before dismantling.

7. Now you can rearrange these four ratchet assemblies to give you a new combination. (Should you want to use numbers that are not in the present combination, your distributor or the Dialoc Corporation can readily supply these.) Remove the ratchet assemblies 469 from the control bar and change your sequence as you desire. Put the ratchet assemblies back on the control bar. When replacing the ratchet assemblies on the control bar, make sure that they go on with the ratchet part of each assembly down and that the first digit of the combination is put on first, then the remainder of the combination in order from the first ratchet assembly outward.

To assemble:

1. Place the four ratchet assemblies 469 on the control bar 726 and, grasping the assembled control bar and the ratchet assemblies, slip them over the center spindle of the primary arm 406. The ratchet assemblies can be guided over the lower mounting pin protruding from the outside plate 820.

2. Care should be taken that the control bar 726 has

Fig. 18-8. The very sophisticated Dialoc 1400. (courtesy Dialoc Corporation of America)

passed through the elongated hole in the bottom of the housing 414.

3. Replace the secondary arm 402 on the primary arm 406. Make sure that the finger 408 and the finger follower spring 412 are in position.

4. If the cam pawl trip 441 has been dislodged during the combination change, it can now be replaced by inserting it into the housing with the small protruding end down. The blade on this part must be against the ratchet pawls of the ratchet assemblies.

5. Make sure that the secondary arm 402 is flush with the top of the square bar. This allows a spacing of approximately 0.020″ between the secondary arm and the top ratchet assembly.

6. After the secondary arm is properly placed, return control spring 744 to the end of the control bar 726.

7. Replace the housing cover 745. Make sure that the cam pawl trip 441 and the control bar 726 are in their respective holes in the housing cover. A straight pin is helpful in guiding these parts into holes. Be sure reset arm 870 is inside the housing 414.

8. Replace the cover-retaining nuts 484-5. These nuts are to be turned down snugly with the fingers, plus approximately ½ turn with a wrench.

9. The lock should now be operated before replacing on door to insure its proper operation on the new code settings.

Master 1500 Padlock

The Master 1500 combination padlock is covered with heavy-gauge sheet steel, rolled and pressed at the edges. The net result of this somewhat unorthodox construction is a tight-fitting lock and one that is not pried open easily. Disassembly is recommended only as a training exercise. It is not practical to open this lock for repair.

Three wheels are employed, each with a factory-determined number. If you receive the lock with the shackle open you can determine the combination by looking through the shackle hole. Note the dial reading as you align each wheel. If you read the numbers properly and if the wheel gates are in alignment, you should only have to add 11 to the dial readings to get the true combination. If you are slightly off, compensate by adding 10 or 12 to the original readings.

The Master 1500 series may be master keyed (Fig. 18-9). This lock is popular in schools. While the combination can be obtained by manipulation, picking the keyway is faster. Once the lock is open, determine the combination by the method described above.

PUSHBUTTON KEYLESS LOCKS

Combination locks are those devices that do not require a key to open. Although many people think of such things as safes and vaults or perhaps a bicycle padlock requiring a three-digit combination as the only type of locks that fall under the classification "combination locks," this is not true. Any device that does not require a key, but instead relies on a single or series of digits or letters in a certain order to open the mechanism, is classed as a combination locking device.

This concluding section of the chapter will discuss, illustrate, and provide specific details concerning several of the most popular and best keyless locking devices available today. Remember that the products discussed here are not the only types, nor are they representative of the only manufacturers of such products; they are the products that have proven themselves to a wide variety of customers in varied and numerous types of installations. As such, they are recommended for your consideration as

Fig. 18-9. The Master 1525 master keyed combination lock. (courtesy Master Lock Company)

some of the highest quality, professional, and most secure keyless type of combination locking devices on the market today.

KEYLESS LOCKING DEVICES

Keyless entry locks provide exceptional security in a variety of modes. Motels, hotels, various private and government institutions are using these locks with increasing success (Fig. 18-10). The main manufacturer and proponent of keyless locking devices is Simplex Security systems. Their locks are commonly used in a variety of security-related situations where more than one entrance control device is required. As an example, consider the case of a company research and development facility. For added security, a keyless lock is added to each door. At night, then, two locks secure the door. During the day, the main lock is in the unlocked position, but the pushbutton lock is readily available for the use of the R&D employees since they are the only individuals who know the particular combination to gain access.

In motels and hotels, the use of keyless locks has increased security, cut costs, and saved time and manpower. How? By using the Simplex lock, in one example, one motel stopped losing more than 50 keys a month and saved the costs of continually having duplicate keys made and locks rekeyed. Best of all, thefts were almost entirely eliminated.

Time and manpower were also saved with the installation of the Simplex lock. The desk clerk no longer had to hunt a duplicate key or go to a room and unlock it

Fig. 18-10. Simplex 5-pushbutton deadbolt lock. (courtesy Simplex Security Systems, Inc.)

because the key was locked inside. All that is required is to give out the specific room combination, just as was done when the individual checked into the motel.

Once the guest checks out, it requires only three minutes to change the combination; this keeps the room security integrity very high.

The Simplex lock is extremely easy to operate and maintain, despite its very high security. The lock has five pushbuttons arranged in a circle and numbered from one to five. The combinations can be set at random; to operate the lock, the various buttons are pushed in sequence, in unison, or a combination of both to affect entry.

Installation of the lock is also relatively easy. The only tools required are a screwdriver, electric drill and a hole saw set.

Simplex

Before we get into the various installation procedures, let's look at the specific advantage of a Simplex lock so you will understand, as a locksmith, why many businesses and homeowners are installing these locks.

☐ No keys are required. For businesses, this means none are issued or controlled, and there are none to recover when the employee leaves.

☐ They are burglar-proof since there is no keyhole to pick.

☐ There are thousands of possible combination options.

☐ You can change the combination in seconds whenever there is a change in personnel, tenants, or for other reasons of security.

☐ Money is saved since there are no keys or cylinders to be changed or replaced.

☐ The locking units are extremely durable, being wear-tested for the equivalent of 30 years of intensive use.

☐ It is an extremely easy-to-install lock, more so than any other existing lock or knob type lockset. It is completely mechanical and requires no electrical wiring or specialized accompanying equipment.

The pushbutton combination lock (Fig. 18-11) comes in several models: the deadlock with a full 1″ deadbolt, the automatic spring latch, the key bypass, and the special security model (for use with Department of Defense or industrial complexes requiring the security necessary for the protection of closed or restricted areas).

The deadlock, model DL, requires only the combination pressed in the right order and the turn knob moved in order to open the lock from the outside. The lock remains in the open position, with the bolt retracted, until either

Fig. 18-11. Deadbolt with a key-bypass. (courtesy Simplex Security Systems, Inc.)

the inside lever or outside knob is turned to throw the bolt into the locked position.

The automatic spring latch, model NL, locks automatically. It opens with a finger lever from the inside. A thumb slide on the inside holds the retracted latch bolt in the unlocked position, if desired (Fig. 18-12). Lacking the latch holdback is a variation, the model NL-A. A deadlocking latch can also be ordered for these units; this will

prevent prying the latch back on the closed door. The latchbolt is reversible without affecting the deadlatching function.

The key bypass models (Fig. 18-13), DL-M and NL-M, are the DL and NL models with an optional key cylinder. They provide the flexibility of a master key system and the security advantages of a combination lock. Employees, tenants, and others use the combination. The key ensures entry regardless of the combination setting used, and is given only to selected and designated management personnel. This eliminates the need to maintain a central listing of combinations for reference use by top management personnel. It is completely safe as there is no widespread proliferation of keys, and key distribution can be severely restricted.

The key bypass model is extremely popular in facilities working for and with the Department of Defense and other government agencies, due to its excellent security advantages.

The special security model, NL-A-200-S, is designed for complete compliance with the Industrial Security Manual of the Department of Defense. The model is an approved substitute for the expensive guard forces in the protection of closed and restricted areas in contractor facilities. It has all the features of the model NL automatic spring latch, plus:

☐ A face plate shield to prevent observations of pushbut-

Fig. 18-12. Finger lever and thumb slide for the Simplex, Mode DL lock; inside view of door and back of lock. (courtesy Simplex Security Systems, Inc.)

Fig. 18-13. The key bypass is also available with the standard latchbolt model Simplex lock. (courtesy Simplex Security Systems, Inc.)

Fig. 18-14. Face-plate shields will prevent the observation of the combination code by unauthorized individuals. (courtesy Simplex Security Systems, Inc.)

ton operation and maintain the secrecy of the combination (Fig. 18-14).

☐ The combination change access can be padlocked to ensure authorized combination changes only.

☐ The latch holdback feature is eliminated so the lock can never be kept open. It is in the locked condition at all times.

All of the various Simplex models have the same basic three-piece self-aligning assembly and installs easily and quickly. *Note:* All Simplex door locks require a minimum flat surface of 3½″ on the inside of the door to mount.

Simplified, the procedures for installation are:

Wood and door frame installation: Just drill two holes, ¾″ and 1⅝″, a distance of 2⅝″ from the door edge. Key bypass models will require an additional ⅞″ key

Fig. 18-15. Standard and optional strikes. (courtesy Simplex Security Systems, Inc.)

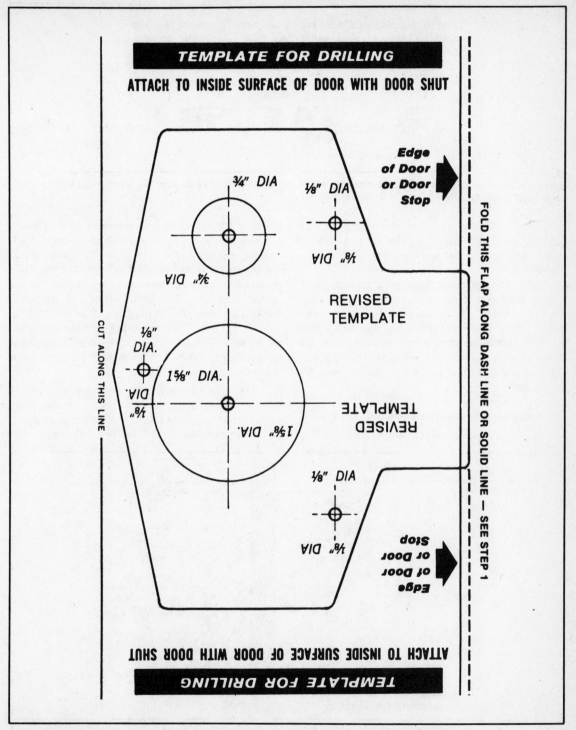

Fig. 18-16. Drilling template for the 100 and 200 series (DL /NL) model locks. (courtesy Simplex Security Systems, Inc.)

Fold template on **Dash Line** for additional clearance required on outswing doors or when using Part #100 Riser Plate on inswing doors.

Fold template on **Solid Line** on inswing doors when **not** using Riser Plate.

OUTSWING

INSWING

Fig. 18-17. This visual provides for the proper determination of which template line will be used for positioning the lock properly to the door. (courtesy Simplex Security Systems, Inc.)

cylinder hole. All parts are provided for a complete wood door and frame installation. Included are two styles of strikes (standard) to suit any wood door frame ("A" in Fig. 18-15).

Metal door and frame installation: An adapter kit containing a special surface-mounted strike ("B" in Fig. 18-15) is available. The special strike is fastened to the inside surface of the metal frame and eliminates the need to cut or mortise the frame for the standard strike. The adjustable riser plate raises the lock housing to ensure the proper alignment of the latch or deadbolt with the special strike.

Installation Instructions—Models DL and NL:

1. Using the template provided with the lock (Fig.

18-16), cut the template from the sheet and fold on the printed line as shown by the arrow. With the door closed, tape the template to the inside of the door with the fold line aligned with the edge of the door (or door stop if one is present). The same template applies to both inswing and outswing doors (Fig. 18-17). Using a center punch, mark the centers of the two large and three small holes.

2. Using a ⅛″ diameter drill, drill the centers of the ¾″ and 1⅝″ diameter holes all the way through to the front surface of the door. Drill the three mounting screw holes approximately ½″ deep.

With a 1⅝″ diameter hole saw, open up the larger hole by drilling from both the outside and the inside surfaces, meeting in the middle. Follow the same procedure with a ¾″ diameter drill for the smaller hole. *Note:*

For all NL Latch type locks clearance between lock housing and strike must not exceed 1/8″ to ensure deadlatching.

Fig. 18-18. Lock housing and holding bracket positioning for proper lock installation. (courtesy Simplex Security Systems, Inc.)

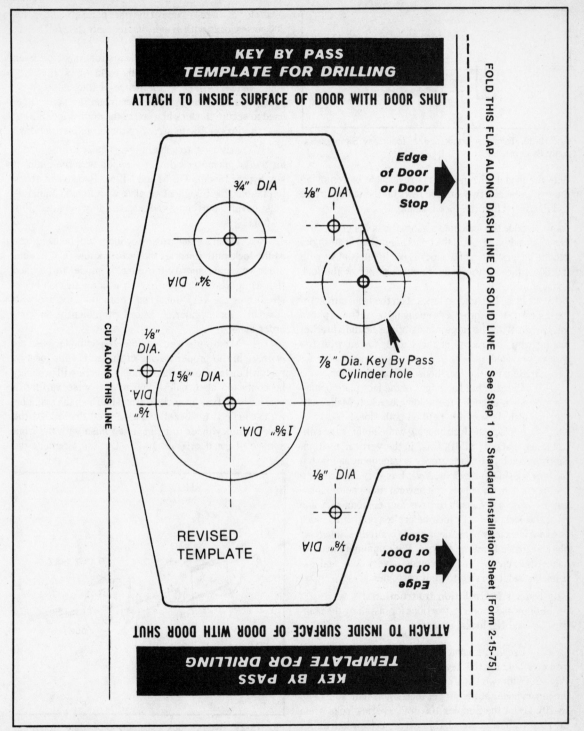

Fig. 18-19. Drilling template for the key bypass feature. (courtesy Simplex Security Systems, Inc.)

Fig. 18-20. Tailpiece break points. (courtesy Simplex Security Systems, Inc.)

With the model NL, the latch is reversible by removing the two large and two small screws on the undersurface and disassembling the lock housing.

3. Slide the lock housing into the holes in the door as shown in Fig. 18-18. If the holes are drilled at an angle, and the undersurface of the lock is not in full contact with the door, file out the holes to correct. Screw the lock housing to the door.

4. Place the holding bracket on the front surface of the door with the slotted legs engaging the aligning pin on the barrel. With the large radius of the holding bracket mating with the radius of the barrel, fasten with the screws provided.

5. Remove the lock housing from the door.

6. Slide the holding edges of the face plate behind the formed up edges of the holding bracket. Make sure the face plate is securely held on both sides.

7. Replace the lock housing while holding the control knob, marked "SIMPLEX," in the vertical position. Turn the knob to ensure proper engagement and fasten the lock housing to the door. *Note:* It is very important to try the combination of the unit several times before locking. It is also a necessity to have your customer try and open the lock several times before leaving. If the lock does not fit exactly to the door, there is an adjustable riser plate available from the factory. With minimum maintenance, this unit will now provide a very long and extremely useful service to your customer.

Key Bypass Installation Instructions:

Follow the above instructions for installing the standard DL and NL locks, except:

1. Use the template (Fig. 18-19) for positioning ⅞″ diameter hole for the key lock cylinder.

2. With two pairs of pliers, break the tail piece of the key cylinder at the proper location as indicated (Fig. 18-20). (1) for the Simplex 100 or 200 series locks when the riser plate is not used; (2) for 100 series when either ¼″ or ⅜″ side of the riser plate is used, or with the 200

series locks when ¼″ side of the riser plate is used; (3) for 200 series locks with ⅜ side of riser plate, use full length of tail piece.

3. After step 6 of the basic instructions have been taken, assemble the key lock cylinder in the ⅞″ diameter hole. *Note:* The hole in the flange of the cup must be positioned over the ⅛″ diameter mounting screw hole used to secure the lock housing to the door (Fig. 18-21).

4. Proceed with step 7 of the basic instructions.

You may wish to substitute a standard rim cylinder for the ⅞″ diameter cylinder supplied with the Simplex key bypass models DL-M and NL-M. Because of space limitations, the Simplex face plate with Roman numerals must cover a portion of the standard key cylinder (Fig. 18-22).

1. When a standard key cylinder will be used, you as the locksmith must let the factory know this; in your order you must specify the Simplex model to be used (DL-M or NL-M, 200 series, and whatever finish is desired) plus the very important notation "...for use with standard rim cylinder..." placed prominently on your order blank.

2. Make sure you get the *rim* cylinder, *not the mortise*. In the proper keying of the unit, your local or regional distributor or other supply source will probably be contacted also, unless you have a wide variety of various lock parts in stock. You must furnish the cylinder. An escutcheon (or bezel) must be used that allows the nose of the cylinder to be mounted flush with the front surface of the door (Fig. 18-23). Cut the length of the

Fig. 18-21. Bypass lock assembly procedure. (courtesy Simplex Security Systems, Inc.)

Fig. 18-22. Simplex lock using a standard sized rim cylinder. Note how the Simplex overlaps a portion of the cylinder. (courtesy Simplex Security Systems, Inc.)

Fig. 18-24. Arrow indicates lock housing screw that must be attached to the cylinder back plate. (courtesy Simplex Security Systems, Inc.)

tailpiece to suit. Cut or grind the width of the tailpiece to 3/16″ to fit the tubular coupler on the Simplex lock.

3. Use the Simplex key bypass template (Simplex Form 2-15-65A) for drilling. Use the center of the ⅞″ diameter hole for drilling the appropriate size hole for the standard rim cylinder and escutcheon. Assemble the cylinder, escutcheon and back plate to the door *before* mounting the Simplex lock.

4. Now install the Simplex by following the standard installation instruction procedures. The Simplex face plate (with Roman numerals) and holding bracket will cover part of the cylinder. It may be necessary to notch the edge of the escutcheon to clear one of the two screws used to fasten the holding bracket to the front surface of the door. One of the two lock housing screws nearest the door edge must be fastened to the back plate of the rim cylinder (Fig. 18-24). Using the hole in the lock

housing as a guide, drill a .161″ diameter hole (#20 size drill) in the back plate for the #10 sheet metal screw.

After the Simplex lock is installed, you may have a service and maintenance contract for the numerous Simplex locks in the building. During the course of your locksmithing career you will be called upon many times to carry out your portions of the service contract with the building manager. Among the several procedures most common for the Simplex lock, you may be changing the combination for one or many of the locks or be required to disassemble the lock for one reason or another. You will also be expected to be able to find an unknown combination. The procedures for these various parts of the locksmith's job are detailed below.

Combination Changing:

With the door in the open position and the lock in the

Fig. 18-23. Standard lock cylinder positions with the door front surface. (courtesy Simplex Security Systems, Inc.)

Fig. 18-25. Use the Allen wrench to remove the access screw in order to change the combination of the lock. (courtesy Simplex Security Systems, Inc.)

Fig. 18-26. Simplex Unican 1000 Series Heavy Duty Lockset. (courtesy Simplex Security Systems, Inc.)

☐ Built-in security with thousands of possible combinations.
☐ No keyway to pick.
☐ Weatherproofing which allows for all types of installations.
☐ A ¾" deadlocking latch with a 2¾" backset.
☐ Completely mechanical.
☐ No electrical wiring.
☐ One-hand operation.

The unit is 7¾" high, 3" wide, and 3 13/16" deep. Latch front is 1⅛"×2¼" with a backset of 2¾". The Unican lock will fit doors with thicknesses ranging from 1⅜" to 2¼".

A key bypass option is available, mainly for master keying purposes, using the Best or Falcon removable core cylinders in the outside knob (Fig. 18-27). The removable core cylinders are not available with the lock when ordered, but must be obtained separately. On a special order basis, the Russwin/Corbin removable core capability can be obtained.

The second optional feature is the passage set func-

locked position, push the existing combination. The hexagon screw at the top of the lock housing (Fig. 18-25) is then removed with an Allen wrench. After removing the screws, the wrench is inserted into the hole and a red button inside is depressed. The wrench is then removed.

The front control knob (below the pushbuttons and marked "Simplex") should then be turned to the left. This removes the existing combination.

The pushbuttons on the front are now depressed in the sequence desired for the new combination. (This can be up to five single numbers or two at a time simultaneously.)

It is advisable to write down the combination sequence immediately and present it to your customer for safekeeping. The front knob marked Simplex should now be turned to the right. The new combination is now set in the lock. The combination should be tried several times with the door in the open position before attempting it with the door in the closed position and the door, for all intents and purposes, locked. (At this point, stand inside the door and let the customer operate the combination at least once from the outside.)

The Simplex Unican 1000 series lock (Fig. 18-26) is a heavy-duty lockset ideal for securing high traffic areas in commercial, institutional, and industrial buildings, apartment lobbies, hotels, motels, luxury homes, and restricted areas in commerce and industry. Like other simplex high quality products, the Unican 1000 has:

Fig. 18-27. Unican 1000 with key bypass feature. (courtesy Simplex Security Systems, Inc.)

tion. When activated by an inside thumb turn or key cylinder, the outside knob can be set to open the lock without the need for using the pushbutton combination (Fig. 18-28). This is designed for use in offices, etc., when the door is to be left unlocked for an extended period of time.

Both 9″ and 14″ high "mag" type filler plates are available to cover up previous lock installation holes on 1¾″ doors. These three-sided steel plates wrap around the door extending 5½″ on both the outside and inside surfaces.

Unican 1000 Series Installation.

The Unican 1000 uses an ASA 161 cutout (2⅛″ diameter through hole with a 2¾″ backset) as found with most other heavy-duty cylindrical locksets (Fig. 18-29). In addition, two ¼″ diameter holes are required to bolt the top of the lock case to the back side of the door, and a 1″ diameter hole for the key operated combination change mechanism to operate from the back side as well. The

Unican can be installed in wood or metal doors from 1⅜″ to 2¼″ thick.

In addition to the special tool provided with each lock, you, as the installer, will also need:

☐ Drill (½″ diameter electrical drill preferred).
☐ 2⅛″ diameter hole saw (for metal doors) or spur bit (for metal doors).
☐ 1″ diameter hole saw or spur bit.
☐ ¼″ diameter drill.
☐ Two wood chisels (wood doors) ¼″ and 1″ wide.
☐ Phillips and standard screwdrivers.
☐ Center punch and hammer.
☐ Standard 12″ ruler.

Before starting the actual installation procedures, let's look at a portion of the Unican three-part 1000 template that will be used. (Figure 18-30 shows the template and proper dimensions with Fig. 18-31 the door edge portion of the template.) Figure 18-32 provides the

Fig. 18-28. Key bypass or thumb turn can be employed with the unit. (courtesy Simplex Security Systems, Inc.)

267

Fig. 18-29. Exploded view of Unican 1000 positioning for a wood door. (courtesy Simplex Security Systems, Inc.)

information and specifications for the latch cutout required. *Caution:* When using this type template, always apply the template and drill from the outside, being sure to compensate for the door bevel, if any (Fig. 18-33).

1. First, make all necessary corrections to ensure the door is properly hung in the frame. Apply the template to the high edge of the door bevel and tape it in place. Should a metal door frame preposition the strike location, be sure to position the template to center the 1″ diameter latchbolt hole in the strike cutout in the frame. With a center punch and hammer, mark the centers of all required holes.

2. With the ¼″ drill, drill the centers of the 2⅛″ and 1″ holes all the way through the door. Should you wish to center mark the hole on the other side of the door (the low side of the bevel) using the template, be sure to move the "fold" line to shorten the backset to compensate for the bevel. Laying a carpenter's square on the high edge of the bevel will show how much compensation to make.

It is a good practice to bore the 2⅛″ and 1″ diameter

holes halfway in from both the front and back surfaces of the door.

If the door was originally prepared for a 2⅜″ backset, you can file out the 2⅛″ hole by an additional ⅜″ as the lock case and inside rose are wide enough to cover.

Be sure to drill the 1″ latchbolt hole on the door edge straight and level to prevent binding of the latch assembly. Carefully drill the two ¼″ mounting holes through the door. It cannot be overemphasized that a drill jig is of great assistance in keeping your mounting hole straight and true.

Care must also be taken to allow clearance for the hinge that operates the antifriction device on the Unican latch. The latchbolt hole must be relieved adjacent to the antifriction trigger to permit the hinge to rotate freely and prevent the binding of the latch.

3. It is imperative for a proper Unican 1000 installation to have a freely operating latch properly positioned to the strike. Excessive binding or dragging of the latch can cause the front knob clutch to "freewheel," preventing the latch withdrawal. Also, the deadlatching plunger must be held depressed by the strike plate to be operable. The gap between the door and the frame must never

Fig. 18-30. Unican 1000 door template. (courtesy Simplex Security Systems, Inc.)

Fig. 18-31. Unican 1000 template portion for marking the door thickness. (courtesy Simplex Security Systems, Inc.)

Fig. 18-33. Details of the bevel compensation for the Unican 1000. (courtesy Simplex Security Systems, Inc.)

be more than a ¼" to ensure proper deadlatching. If door silencers are present, allowances must be made in positioning the strike.

To properly position the strike, install the latch assembly into the edge of the door. On wooden doors, mark around the faceplate of the latch and mortise in with a chisel. Remove 5/16" minimum for antifriction trigger hinge and 6/32" of material for the faceplate assembly. When inserted, the latch assembly should be flush with the nose of the door. Close the door to the point where the latch touches the door frame and mark the frame at the top and bottom lines previously scribed. Mark the screw holes. When the door is closed, the screw holes of the strike should be even with the screw holes of the latch plate. Cut or mortise the frame to a minimum depth of ¾" to ensure a full throw of the latch when the door is closed. Combinations cannot be cleared or changed if the full travel of the latch is impaired. The Unican 1000 comes factory packed with a 2¾" strike (and strike box for metal frames). An ASA 4⅞" strike is also available as a accessory from the manufacturer. Do not use a strike with a lip radius different than that supplied by the manufacturer. A short radius on the lip can damage the latch assembly.

4. Unless otherwise specified in your order, all Unican 1000 series locks are factory assembled for left-hand operation. To change it to right-hand operation, remove the back plate containing the cylindrical unit from the lock case by removing the Phillips head screws. Remove the four Phillips head screws holding the cylindrical unit at the back plate (Fig. 18-34) and rotate the cylindrical unit so that the cutout for the latch faces in the opposite direction. Reattach the cylindrical unit using the four screws (with two lock washers for each screw), making sure the unit is centrally positioned to the hole in the back plate and rotates freely when the spindle is turned. To prevent internal jamming, two lockwashers must be used with each screw. Remount the back plate assembly to the lock case.

Fig. 18-32. Unican 1000 latch cutout specifications. (courtesy Simplex Security Systems, Inc.)

Fig. 18-34. The four screws must be removed so that the cylinder can be rotated, if necessary. (courtesy Simplex Security Systems, Inc.)

5. All locks are factory set for a 1¾″ thick door. To adjust for a 1⅜″ to 1½″ door, remove the back plate and cylindrical unit assembly from the lock case and remove the spacer that is positioned between the cylindrical unit and the back plate (Fig. 18-35). Reassemble the cylindrical unit with the shorter (6-32″×¼″) screws provided with the unit. Insert the roll pin in the front knob shaft one hole closer to the front knob (from position #2 to #3 in Fig. 18-35). Again, be sure that there are two lock-

washers under each screwhead joining the cylindrical unit to the back plate.

To adjust for 2 to 2¼″ doors follow the above instructions, but add the extra spacer provided (lip edge to lip edge) to the one already assembled to the lock. Use the longer (#6-32×¾″) screws provided and insert the pin in the shaft in the hole furthest from the front knob (from position #2 to #1 in the shaft in the hole furthest from the front knob (from position #2 to #1 in Fig. 18-35). When remounting the cylindrical unit, be sure it is central to the hole in the back plate and turns freely. Double check to ensure the two lockwashers are under each screwhead.

6. Hold the latch assembly in the edge of the door and slide the lock case in from the front. Depress the latch slightly to engage the "T" (marked "A" in Fig. 18-36) in the lock retracting shoe (marked "B"). Viewing the unit from the back side of the 2⅛″ hole, be sure the curved lips of the latch housing slide within the opening of the cylindrical case ("C") in Fig. 18-36. Hold the lock case to the front of the door with one hand and insert the two top mounting screws from the back. Finger-tighten both screws, to start, to ensure the holes were properly positioned. If you are not able to start the screws into the back of the lock case by finger tightening, the mounting holes were not positioned correctly. Forcing can cause the threads to be stripped. To correct this condition, remove the lock case and file out the holes to ensure proper alignment of each to the nuts on the back plate.

7. Screw on the inside rose and tighten with the spanner wrench. Slide the knob onto the inside sleeve, depressing the knob catch spring by pushing the pointed tip of the wrench through the hole in the side of the knob

Fig. 18-35. Backplate assembly and cylinder unit must be removed in order to adjust the lock for door thickness variations. (courtesy Simplex Security Systems, Inc.)

Fig. 18-36. Latch assembly with critical portions indicated; these three points (A-B-C) must be worked in exact order to ensure proper lock installation. (courtesy Simplex Security Systems, Inc.)

271

Fig. 18-37. The trim plate *must* be positioned horizontally over the hole in the door. (courtesy Simplex Security Systems, Inc.)

the front knob fully clockwise and then release. Push the existing combination and then release the buttons. The buttons should never be held depressed during lock opening. Insert the flat blade end of the special tool into the slots of the combination change sleeve and rotate clockwise to stop, but do not force. Then rotate counterclockwise to stop and remove the tool. The combination change sleeve will rotate easily when the correct combination is pushed. It will not rotate with an incorrect combination, and forcing can damage the mechanism.

Now turn the front knob fully clockwise to stop and

collar. When the knob is pushed into the stop, remove the tool and make sure the knob is securely fastened.

8. The trim plate is mounted horizontally over the 1″ diameter combination change hole and the two mounting screws (Fig. 18-37). With the center punch, indent the door slightly for the two protrusions on the plate used to prevent rotation. Cut the lock screw to length as required. Insert the control key and keep turning clockwise to thread the lock screw into the combination change sleeve assembly on the back plate of the lock case. Tighten the screw until the plate is snug to the door, but do not overtighten. The key can only be removed in a vertical or horizontal position. Overtightening may cause the lock screw to jam in the change sleeve, requiring the removal of the screw with your needle-nosed pliers.

9. All locks are factory set for the same combination: buttons #2 and #4 are pushed together, and then #3 individually. For security reasons, you should change the combination immediately, using as few as one button, or as many as five. When the buttons are set for pushing at the same time with other buttons, it is a different combination than if the buttons are set to be pushed individually (i.e., 1, 2, 3 individually pushed buttons is a different combination than if 1, 2, 3 are pushed at the same time). The same button cannot be used more than once in any given combination. If your Unican has a passage set function, be sure it is not in use when the combination code is changed.

To set the combination, the door must be open. Remove the inside trim plate with the control key. Turn

Fig. 18-38. M.A.G. filler plate. (courtesy Simplex Security Systems, Inc., and M.A.G. Engineering and Mfg., Inc.)

Fig. 18-39. Exploded view of the Unican 1000 Lock unit. (courtesy Simplex Security Systems, Inc.)

release. Turn only once at this stage of the combination change. Depress the buttons firmly and deliberately in the sequence desired for the new combination. Record the combination at once so it will not be forgotten. Turn the front knob fully clockwise to stop to set in the new combination. Try the new combination before closing the door and be sure you have properly recorded it. *Note:* If the Unican will open without pushing a code, you are in the "0" combination mode because the front knob was turned more than once, or out of sequence. In such a case, repeat all the steps in the change procedure except pressing the existing (old) combination. It is very important that the front knob be turned fully *clockwise to stop*

when called for in the combination change procedure to ensure that the old code is completely cleared out and the new code is completely set in. Full rotation will be impaired by any of the following:

☐ Using a latchbolt assembly not supplied by the manufacturer (especially one with a shorter backset or shorter throw).

☐ A shallow strike box that prevents the full ¾" throw of the latch when the door is shut.

☐ Any binding of the latch assembly in the door or in the strike caused by improper door or frame preparation or by door sag and/or buckling.

273

ITEM NO.	PART NUMBER	DESCRIPTION	PER UNIT
1	200024	Push Buttons	5
2	200026	Bushing, Shaft	1
3	200079	Spring, Nut	1
4	200080	See Assembly Chart	1
5	201006	Link, Combination Chamber	1
6	201016	Spring, Balance	1
7	201018	Spring Double Wave	1
8	201019	Knob Return Spring	1
9	201037	Inside Trim Plate Only	1
10	201038	Pin, Linkage	1
11	201040	Stop Plate	1
12	201048	See Assembly Chart	1
13	201049	Front Plate	1
14	201050	See Assembly Chart	1
15	201093	See Assembly Chart	1
16	201102	Spacer Door Thickness	2
17	201144	Latch	1
18	201155	See Assembly Chart	1
19	201218	Drive Insert	1
20	201267	See Assembly Chart	1
21	201272	Shoe	1
22	201275	Inside Sleeve	1
23	201276	Shoe Retainer and Bridge	1
24	201277	Cover for Shoe Housing	1
25	201278	Installation Wrench	1
26	201280	Rose	1
27	201281	Rose Reinforcing Plate	1
28	201282	Thread Ring	1
29	201284	Knob Standard	2
30	201287	Strike Box	1
31	201288	Strike Plate	1
32	201289	Shoe Retainer Cap	1
33	201290	O Ring 1 in. I.D.	1
34	201291	Screw, Philips Combination No. 8	4
35	201293	Clip	3
36	201298	Key DF 59 and Key Ring	2
37	201301	See Assembly Chart	1
38	201303	See Assembly Chart	1
39	201304	Clutch Backing Plate	1
40	201335	Combination Change Stud	1
41	201336	Combination Change Sleeve	1
42	201343	Front Knob Retainer Clip	1
43	201344	Inside Knob Retainer Clip	1
44	201345	See Assembly Chart	1
45	201350	Chamber Release	1
46	201351	Chamber Release Cam	1
47	201352	Bushing, Passage Set	1
48	201358	Connecting Bar	1
49	201362	Washer	1
50	201365	Knob Insert Best/Falcon	1
51	201366	Shoe Spring	2
52	201367	Turn Knob	1
53	201397	See Assembly Chart	1
54	201398	See Assembly Chart	1
55	201414	Knob Insert Russwin/Corbin	1
56	201416	See Assembly Chart	1
57	201417	See Assembly Chart	1
58	201422	Key DF 5 for Passage Control	2
59	201425	Shaft Self-Aligning Bushing 1000-2,4 & 6	1
60	201426	See Assembly Chart	1
61	201427	See Assembly Chart	1
62	201429	Combination Chamber	1
63	201430	See Assembly Chart	1
64	201431	See Assembly Chart	1
65	201432	See Assembly Chart	1
66	201433	See Assembly Chart	1
67	201434	See Assembly Chart	1
68	201435	See Assembly Chart	1
69	201436	Cross Pin .075 Dia.	1
70	201437	Cross Pin .100 Dia.	1
71	201438	Back Plate	1
72	201439	See Assembly Chart	1
73	201440	Retainer Ring 3/8 I.D.	1
74	201441	Retainer Ring 1 1/4 I.D.	1
75	201442	Screw Stud 1/4-20 X 1 5/8	1
76	201443	Screw 8-32 X 2 1/2	2
77	201444	Screw Philips 6-32 X 3/8 F.H.M.S.	6
78	201445	Split Washer No. 8	17
79	201446	Screw Philips 6-32 X 1/4 R.H.M.S.	4
80	201447	Screw Philips 6-32 X 5/16 R.H.M.S.	2
81	201448	Screw Philips 6-32 X 3/8 R.H.M.S.	3
82	201449	Screw Philips 8-32 X 1/4 R.H.M.S.	4
83	201450	Screw Philips 8-32 X 1/2 R.H.M.S.	4
84	201451	Screw Philips 8-32 X 3/4 R.H.M.S.	4
85	201490	See Assembly Chart	1
86	201496	See Assembly Chart	1
87	201497	See Assembly Chart	1

ASSEMBLY CHART

ASS'Y NO.	INCLUDING ITEM NUMBERS	ASSEMBLY DESCRIPTION
200080	4, 36	Control Lock Assembly Combination Change (DF-59)
201048	19, 21, 22, 23, 24, 28, 32, 33, 43, 51, 74, 78, 79	Cylindrical Drive Unit Assembly
201050	14, 69, 70	Shaft Assembly Standard
201093	15, 50	Knob Assembly Key Override for Best/Falcon
201155	3, 4, 9, 36, 75	Inside Trim Plate Assembly
201267	20, 69, 70	Shaft Assembly, Key Override
201301	16, 19, 21, 22, 23, 24, 32, 33, 40, 41, 43, 49, 51, 71, 73, 74, 78, 83	Back Plate Assembly
201303	26, 27, 28	Rose Assembly
201345	22, 32, 33, 43	Inside Drive Sleeve Assembly
201397	9, 34, 48, 58, 70, 86	Key Actuator Assembly for Passage Set
201398	9, 34, 48, 52. 70	Turn Knob Actuator Assembly for Passage Set
201416	55, 56	Knob Assembly Key Override for Corbin/Russwin
201417	20, 42, 55, 56, 59, 65	Conversion Assembly 1000-1 to 1000-2C
201426	1, 2, 8, 13, 14, 29, 42, 62, 63, 66, 67, 78, 80, 81	Front Plate Assembly 1000-1
201427	1, 2, 8, 13, 15, 20, 42, 50, 59, 62, 63, 64, 67, 78, 80, 81	Front Plate Assembly 1000-2B
201430	6, 7, 63	Clutch Sub-Assembly
201431	11, 64	Sleeve Drive Assembly Key Override for Best/Falcon
201432	11, 65	Sleeve Drive Assembly Key Override for Corbin/Russwin
201433	11, 66	Sleeve Drive Assembly Standard
201434	39, 67	Clutch Cover Assembly
201435	5, 10, 35	Link Assembly
201439	40, 41, 49, 71, 73	Back Plate Sub-Assembly
201490	15, 20, 42, 56, 59, 64	Conversion Assembly 1000-1 to 1000-2B
201496	58, 86	Control Assembly for Passage Set (DF-5)
201497	1, 2, 8, 13, 15, 20, 42, 55, 56, 59, 62, 63, 65, 67, 78, 80, 81	Front Plate Assembly 1000-2C

Fig. 18-40. Parts identification and assembly chart for the Unican 1000 lock unit. (courtesy Simplex Security Systems, Inc.)

Fig. 18-41. Preso-Matic keyless pushbutton deadbolt combination lock. (courtesy Preso-Matic Lock Co., Inc.)

Further installation tips:

☐ M.A.G. type filler plates (Fig. 18-38) in 9″ and 14″ lengths are available for covering up existing mortise lock preparation or badly butchered doors. These can be obtained for 1¾″ doors only. They are handy to use for an installation template as well.

☐ Do not allow either the inside or the outside knob to hit against the wall. A floor mounted door stop should be used to prevent damage to the factory adjusted clutch mechanism behind the knob. The manufacturer's one year warranties are voided if the knobs are allowed to hit. If there is a door closer, be sure that it is properly adjusted for the Unican.

☐ Due to its unique design, clockwise rotation of the front knob performs two functions. It clears out any incorrect combination attempts and reactivates any buttons previously pushed, and it withdraws the latch bolt if the combination has been pushed properly.

☐ When operating the lock, *never* hold the buttons depressed while turning the front knob since damage to the unit can result. Just depress the buttons until they stop and then release them.

☐ Proper installation and operation of the Unican will

result in many years of satisfied customer usage. Just remember that it starts with a good installation.

Servicing of the Unican 1000 series may be required. In the majority of instances, servicing will be the result of customer mishandling and abuse of the unit. Once in a very great while the unit itself may malfunction for some unknown reason. It is at this time that you require some basic information concerning the various parts of the Unican 1000 unit and illustrative material for the proper assembly or disassembly. Figure 18-39 is an exploded view of the Unican 1000. Figure 18-40 is a parts item description and the assembly number and parts.

All Simplex auxiliary locks are shipped with the following combinations: #2 and #4 pushed at the same time, then #3 pushed singly.

Some people forget how to operate this lock (I, for one, did, for several weeks at a time) even though the procedure is very simple:

1. Turn the control knob to the left to activate the buttons.
2. Press the correct buttons in the proper order.
3. Release the last button(s) before turning the control knob. Turn the control knob to the right to open.
4. To relock, turn the control knob to the left. Model NL locks automatically.

Some customers will look at the lock they contemplate purchasing and wonder about the control knob. Since it operates the bolt mechanism, if it is removed, can the lock be opened?

The answer is a resounding no. The front control knob cannot be forced to open the lock since it is connected to the lock housing by a friction clutch. If the knob has been forced, it will be at an angle and can be turned back to the vertical position by hand or with a pair of pliers without damaging the lock.

Preso-Matic

The Preso-Matic keyless pushbutton combination lock comes in either a deadbolt version (Fig. 18-41) or a deadlatch version (Fig. 18-42). This is another completely mechanically operated lock; no electricity is required. From the outside, the customer presses the selected four-digit combination. The deadbolt or deadlatch automatically retracts, opening the door. With the deadlatch model, the lock automatically locks each time the door is closed.

With several models available, each customer can

Fig. 18-42. Deadlatch version of the same lock. (courtesy Preso-Matic Lock Co., Inc.)

Fig. 18-43. Instant exit feature and also locking ability with just a touch of a button. (courtesy Preso-Matic Lock Co., Inc.)

Fig. 18-44. Combination slides used with the lock unit. (courtesy Preso-Matic Lock Co., Inc.)

have a version with the appropriate options to meet his special need. All models comes with an instant-exit feature that requires only the depressing of an unlock button on the door inside (Fig. 18-43). Also available is a night-latch button, which provides an added measure of protection for the home or apartment owner or the business person working late at night. It means that nobody is going to interrupt (or surprise!) him at work or at home.

For businesses that rely upon electric strikes operated from remote locations, the unit can also be used. The Trine model 012 electric strike unit is given a unique configuration by the Preso-Matic factory to accept the deadlatch bolt.

The Preso-Matic lock has several combination possibilities:

☐ Four-digit combination.
☐ Seven-digit combination.
☐ Master combination (for use with the four-digit combination version).

The four-digit combination provides for a maximum of 10,000 possible combinations; the seven-digit combination can give up to ten million possible combinations. The Master Combination (MC) provides for a master combination in addition to the individual four-digit combination for units. In this way, all locks of a given system will open on one combination (the MC) which is a six-digit number, but each unit within the system can still be opened by its own four-digit combination. The MC combination variations allow for security protection and

Fig. 18-45. Preso-Matic installation template, 2⅜″ backset.

278

AFTER BORING
GOUGE OUT
FOR LUG
AS PER
INSTRUCTIONS

BORE 1" HOLE
PARALLEL TO
SIDES OF
DOOR.

LOCATE
CENTER ON
THIS LINE.

PRESO-MATIC   TEMPLATE

TOP

FOLD ALONG THIS LINE. FOR DOORS HINGED ON RIGHT

2 3/8" BACKSET

1" HOLES

5 3/8"

1 7/8"

PLACE ON OUTSIDE OF DOOR

Fig. 18-46. Preso-Matic installation template, 3″ backset.

279

TEMPLATE

TOP

1" HOLES

5 3/8"

1 7/8"

3" BACKSET

FOLD ALONG THIS LINE. FOR DOORS HINGED ON LEFT

BORE 1" HOLE
PARALLEL TO
SIDES OF
DOOR.

LOCATE
CENTER ON
THIS LINE.

PLACE ON OUTSIDE OF DOOR

Fig. 18-47. Preso-Matic installation, Step 1.

Fig. 18-48. Preso-Matic installation, Step 2.

Fig. 18-49. Preso-Matic installation, Step 2.

MODEL	LOCK TYPE	BOLT PROJECTION	BACKSET	NOTES
8101	Deadbolt	⅝″	2⅜″ (2¾″ also available)	Manual locking. Built-in night latch button. Hardened steel free-turning bolts. Manually locks.
LT8102	Deadbolt	1″	3″	Same
8103	Deadlatch	½″	3″	Night latch button. Locks automatically. Inswinging doors only.
8103A	Deadlatch	½″	3″	Same, except without night latch button.
8200	Deadlatch	½″	3″	Hardened steel latchbolt. Locks automatically. Night button.
8200A	Deadlatch	½″	3″	Same, except without night latch button.

OPTIONS:
1. Change of combination slides for a 7 digit code available.
2. Lock shields.
3. Stay open function; keeps the deadlatch in an unlocked position.
4. Door stop strike plate; deadlatch models only. Case hardened; ideal for double door installation.
5. Extra wide strike plate for metal doors (2½″ × 2½″); case hardened.
6. Heavy duty ANSI electric door opener. 24 Vac. for normally locked units.

Fig. 18-50. Installation layout variations. (courtesy Preso-Matic Lock Co., Inc.)

	A	B	C	D	E	F
Model 8101 2⅜" BACKSET	1⅞	5⅜	1	1½	¹³⁄₁₆	3⅜
Model 8101 2¾" BACKSET	1⅞	5⅜	1	1⅞	¹³⁄₁₆	3¾
Model LT8102 3" BACKSET	1⅞	5⅜	1	2 ¹⁄₁₆	¹³⁄₁₆	3¹⁵⁄₁₆
Models 8103, 8103A, 8200, 8200A	1⅞	5⅜	1	2 ¹⁄₁₆	¹³⁄₁₆	3¹⁵⁄₁₆

Fig. 18-51. Overview of the various lock models. (courtesy Preso-Matic Lock Co., Inc.)

Exploded View Parts List:
1. Large Headed Screw
2. Spring Clips
3. Unlocking Lever
4. Unlocking Lever Pin
5. Bolt Actuator
6. Bolt Housing Casting
7. Multiplier Disc
8. Bolt
9. Bolt Spring
10. Pull Pin
11. Rocker Arm
12. Rocker Arm Pivot Clip
13. Flat Spring
14. Unlock Button
15. Lock Button
16. Inside Cover Plate
17. Long thin spring
18. Pull Plate
19. Combination slide cover plate
20. Combination slides
21. Lock body cover plate
22. Neutralizing spring
23. Number button springs
24. Flat inclined plates on reset button
25. Reset button
26. Number buttons
27. Lock body
28. Deadbolt & Housing

Fig. 18-52. Exploded view of the Preso-Matic keyless combination lock and parts identification. (courtesy Preso-Matic Lock Co., Inc.)

flexibility that are not available with other pushbutton type keyless combination locks. For concerns that require numerous units, the advantages to top management of a master combination are important.

The numbers of the combination for each lock unit are determined by the two "combination slides" (Fig. 18-44) which are inserted into the unit. To change the combination, you simply remove the inside cover plate, turn two spring clips clear of the slot in the lock housing, lift out the combination slides, and slip in new ones.

All the Preso-Matic locks are carefully designed and engineered to aid you in quick, easy, trouble-free installation. Actually, there are only three basic steps:

1. Bore the opening for the lock and latch (Fig. 18-45) using the required template that comes with the unit (Figs. 18-46, 18-47).

2. Mortise the door edge for the bolt face and insert the bolt housing and tighten the screws (Fig. 18-48).

3. Insert the lock and tighten only one screw, connecting the bolt housing to the lock and making the unit operable (Fig. 18-49). Attach the cover plate on the inside of the door. Installation is complete.

More specific installation procedures will be given further on. It must be noted that for volume installations in large construction projects or numerous prefabricated doors, you should use the Preso-Matic round face drive-in bolt which installs in seconds and saves time.

Three basic installation layouts are available, as shown in Fig. 18-50. Figure 18-51 is a quick summary of the various lock models. Figure 18-52 provides an exploded view of the lock and part identification.

Chapter 19

Electrical Access and Exit Control Systems

Electrically operated release latch strikes and locks are easily installed in place of standard units to provide remote controlled access for a door. These units are not new to business and industry and really shouldn't be that new to the locksmith—but they are. Why? Because, except for a few instances in large cities or industrial regions, the average locksmith doesn't get involved with such units. Since it is "electrical," the job is passed on to an electrician. By so doing, the locksmith loses business, and valuable business at that.

Throughout this chapter we will be looking at some electrically operated locks and release latch strikes. We will also discuss the circuitry behind the various types of units, where they can be installed, and the potential of such units.

ELECTRIC RELEASE LATCH STRIKES

Electrically actuated release latch strikes are more common than locks, so this subject area will be dealt with first. These units can normally be reversed for either right or left-hand doors. They operate on low voltage and the strikes will fit a hollow jamb channel as shallow as 1.6″ (1.75 for mortise latches) or they can be, if necessary, installed in wooden jambs. Further on in this discussion we'll provide the information necessary for you to consider in unit selection.

In this important area, though, especially since it is a very rapidly-growing one in terms of access control (security), it is important that you recieve that latest and best information available. I contacted the manufacturer considered the nation's leader in the electrical release latch strike field, Adams Rite, in order to obtain the following information. For Adams Rite's assistance I am extremely grateful.

Electrical release strikes are gaining popularity among small and large business concerns; it is evident that the trend will continue to increase at a greater rate during the coming decade. The electrical release strike comes in a variety of sizes (Fig. 19-1) to meet varying needs and requirements.

There are many electrical strikes available on the market, but while all of them meet minimum standards, you should still want to consider the manufacturer and the purpose for which the units were initially designed and developed prior to your purchase of them.

First and foremost in the field is Adams Rite, whose electrical strikes are architecturally sound and designed to add increased security features to areas requiring a certain amount of traffic control. When properly powered

Fig. 19-1. A sampling of the various types and sizes of electric release latch strikes. (courtesy Adams Rite Mfg. Co.)

and installed, these electrical release strikes will provide a very long, maintenance-free life.

The fact that electrical power is used for the operation of the strike in no way makes the unit an electrical appliance, as some people tend to think. With a few exceptions, the electrical circuitry unique to these units are designed to meet the needs of the strike, not the other way around. The hardware specifier (that's you, by the way) should select the strike and expect the electrician (which, again, *could* be you) to supply the power at the point of installation. In a few cases an existing circuit can be used, and thus the strike is selected and installed to it. In selecting strikes, two areas of consideration must be followed: electrical and hardware.

ELECTRICAL CONSIDERATIONS

The first electrical factor is to determine the duty of the unit. Is the operation to be intermittent or continuous? If the door is normally locked and released only momentarily from time to time, it is intermittent. If it is the rare case where the strike is activated (unlocked) for long periods, the duty becomes continuous. A still rarer requirement is reverse action in which the strike is locked only when its current is switched on.

Intermittent duty. For a normal intermittent application, the electrical strike will have to be 24 volts ac. This alternating current call-out gives enough power for almost any entrance, even one with a wind-load situation. Yet the low voltage range is below that requiring U.L. or building code supervision. At this voltage, good, reliable transformers are available.

Continuous duty. When a continuous duty application is required, additional components are added to the unit at the factory.

Reverse action. Long period of unlocking can also be obtained by using a reverse action strike. This might be required to provide the same service as a continuous duty strike, but preserves current because it is on battery dc power. It could also be used to provide a "fail-safe" unlocked door in case of a power failure or building emergency (such as a fire) where quick, guaranteed unlocked doors are required. *Note:* Reverse action strikes, when ordered by the locksmith for installation, must specify the hand of the door on the order. These types are not reversible, as are some other strikes.

Monitoring. If a visual or other signal is required to tell the operator that the electric strike is doing its job properly, then a monitoring strike will be required. This will mean that two sensor/switches are added. One is activated by the latchbolt's penetration of the strike, and the other by the solenoid plunger that blocks the strike's release.

Transformer. Low voltage is necessary for proper electric strike operation, and this is obtained by the use of transformers which are stepped down from the normal ac power voltage to 12, 16, 24, or other lower voltage. Whether or not you perform the electrical side of the installation (when wires are required to be run) three electrical specifiers are necessary:

☐ Input voltage (110—standard household current).
☐ Output voltage (12, 16, 24, etc.—from the transformer unit).
☐ The capacity of the transformer, called volt amps (output voltage times the output amperes).

Skimping on the capacity of the transformer to save a few dollars will underpower the door release and is likely to bring in complaints of "poor hardware." Intermittent duty electric strikes draw from 1½ to 3 amps; a continuous duty model draws less than 1 amp.

In any application where the current draw of the strike must be restricted to less than 1 amp, regardless of the duty, a current limiter which stores electrical energy to relieve high use periods is available.

Wiring. The wiring used must carry the electrical power from the transformer through the actuating switch to the door release. It must be large enough to minimize "frictional" line losses and deliver most of the input from the transformer to the door release. A small diameter garden hose won't provide a full flow of water from the nozzle, particularly if it's a long run; neither will an undersized wire carry the full current!

Electrical troubleshooting. When insufficient electrical power is suspected in a "weak" door release, a simple check can be made. Measure the voltage at the door release while the unit is activated. If the voltage is below that specified on the hardware schedule, the problem is in the circuit—probably an under-capacity transformer, if the current length is short. A long run may mean both a transformer and a wire problem.

The second consideration concerns the hardware itself, and has several points:

Face shape (sectional): With two exceptions, electrical strikes from Adams Rite have a flat face. These exceptions are the two having a radius-type face to match the nose shape of a paired narrow stile glass door.

Face size: The basic Adams Rite size conforms to ANSI standards for strike preparation: 1¼″ × 4⅞″. However, two other sizes are offered to fill or cover existing jamb (or opposing stile) preparation from a previous installation of another unit, such as an MS deadlock strike or a discontinued series electric strike.

Face corners: Strikes are available with two types of corners: round for installation in narrow stile aluminum where preparation is usually done by router, and square corners for punched hollow metal ANSI preparation or for wood mortise installation.

Jamb material: Stikes with vertically mounted solenoids are designed to slip into hollow metal stile sections as shallow as 1.6″. Horizontal solenoids for wood jambs requires an easily bored mortise of 3⅛″ depth.

Previous strike preparation: If the jamb was previously fitted with another, now discontinued strike, such as a Series 002 electrical strike (for example), the new strike unit to be installed—say an Adams Rite 7510—will fit in with only minor alterations. In this example, it would mean increasing the lip cutout ¼″ at each end. In the case where a hollow jamb was originally prepared to receive the bolt from another type deadlock, another latch such as the 4510 series would have to be substituted. In such a case, the radius or flat jamb would cover the old strike cutout.

Lip length: the staidard lip on all basic Adams Rite electric strikes accommodates a 1¾″ thick door which closes flush with the jamb. Where the door/jamb relationship is different, a long lip can be added.

Compatible latch: Certain electric strikes will mate with all latches in a given series. The key-in-knob latches and mortise latches are also made to mate properly with the various electric strikes available.

Handing of the strike unit: All standard operation strikes are unhanded; thus they can be installed for either a right or a left-hand door. However, reverse action

locked only when the current is on) units requires a specification of the hand when the order for the unit is placed.

To repeat: understand that electrical strikes are not just electrical appliances; they are hardware. In the case of Adams Rite strikes, they are made by hardware people, because they know that the hardware has to be carefully matched to its application.

Also realize that when a unit is ordered for installation, check with the local electrical codes to ensure that you can actually install it. In some jurisdictions, an electrician may be required to do the job.

INSTALLATION AND PROCEDURES

Let's move on to discuss and illustrate some of the procedures required for a variety of strike units.

Series 750 for an Aluminum Jam Installation:

1. Prepare the door jamb per Fig. 19-2.

2. Install the mounting clips to the jamb using 8-32 × ⅜ screws and pressed metal nut. Leave the screws slightly loose to permit easy alignment of the base assembly and clip.

3. Spacers are provided to assure flush final assembly of the face plate and jamb. Add one or more spacers between the jamb and mounting clip when the face plate extends beyond the jamb. When the face plate sets inside the jamb, spacers must be added between the mounting clip and the electric strike case. See detail "A" in Fig. 19-2. To attach spacers to the mounting clip, remove the protective coating from the spacer and press to the desired mounting clip surface. Make sure the clearance hole in the spacer aligns properly with the hole in the mounting clip.

4. Using the wire nuts, connect the wires coming from the unit to the wires coming from the low voltage side of the transformer.

5. Insert the electrical release into the jamb by tipping the solenoid coil into place behind the mounting clip, then drop the unit onto the clips.

6. Attach the unit with the #10 screws and lock washers provided.

7. Attach the face plate using the screws furnished (8-32 × ¼″).

8. Secure the screws holding the mounting clips to the jamb.

Figure 19-2 also shows the dimensions of the electric door release that would be installed in the aluminum jamb.

Fig. 19-2. Jamb specifications and installation details for a radial (curved) aluminum door jamb. (courtesy Adams Rite Mfg. Co.)

Replacing an Obsolete Unit with an Electrical Strike Model 7510 Wood Jamb:

1. Prepare the jamb to the dimensions shown in Fig. 19-3. Care should be taken to mortise cut to clear the solenoid.

2. Using wire nuts provided, connect the wires from the 7510 unit to the wires from the low voltage side of the transformer.

3. Insert the electric release into the jamb. Attach this with #10 flat head wood screws.

Replacing the Obsolete Unit (002 as an example):

1. Remove the 002 unit and disconnect the low voltage wires. Do not remove the existing mounting clips.

2. Enlarge the jamb cutout by removing .250″ from each side.

3. Using the wire nuts, connect the 7510 wires to the low voltage side of the transformer.

4. For a flush installation, it will be necessary to provide a .062 spacer between the mounting surface and the face plate.

5. Insert the electric door release into the jamb and attach with screws.

Figure 19-3 also shows the dimensions of the 7510 electrical door release unit.

The next installation is for an electric strike, model 7520, to meet ANSI specifications. Note the 'molded' strip down the center of the door frame in the illustration. Installation is as follows:

1. Prepare the metal jamb to the dimensions shown in Fig. 9-4.

2. Spacers are provided with the unit to assure the flush mounting of the face plate to the jamb. To use the spacer(s), remove the protective cover and press them to the mounting bracket. Make sure the clearance hole in the spacer aligns with the tapped hole in the bracket.

3. Connect the wires from the unit to the transformer wires.

4. Insert the electric release into the jamb by tipping the solenoid coil into place behind the mounting bracket. Then drop the unit onto the mounting brackets. Attach the unit with the screws furnished.

5. Attach the face plate.

This installation is for an electric strike for a key-in-knob latch, set to ANSI specifications. Here we look at an Adams Rite 7540 model Unit:

1. Prepare the door jamb as shown in Fig. 19-5.

2. Install the mounting clips using the screws and pressed metal nut. Leave the screws slightly loose to permit easy alignment of the case assembly and clip.

3. Use the provided spacers as necessary.

4. Connect the wires from the 7540 unit to the low voltage wires from the transformer.

5. Insert the electric release into the jamb and attach with the #10 screws provided.

6. Attach the face plate and secure the screws holding the mounting clip to the jamb.

This installation is for an electric strike into a wooden jamb:

1. Prepare the wood door jamb to the dimensions indicated in Fig. 19-6.

2. Connect the wires from the unit to the transformer wires.

3. Insert the strike into the jamb and attach with the #10 wood screws.

4. Attach the face plate.

Figure 19-7 provides the technical dimensions of the strike unit to be installed.

CIRCUITRY

Figure 19-8 illustrates the typical wiring diagram for an audible (ac) operating strike unit. Figure 19-9 is the same unit, but with a rectifier used to provide for silent (dc) operation of the strike unit.

Fig. 19-3. Dimensions and installation details for unit to be mounted in a wood frame. (courtesy Adams Rite Mfg. Co.)

Fig. 19-4. Jamb preparation and dimension data for a metal frame that has a molded strip down the frame center. (courtesy Adams Rite Mfg. Co.)

You may have an instance where the installation requires a monitoring signal operation. In this case, Fig. 19-10 provides such a typical wiring diagram. Many possible monitoring arrangements can be had: the green light indicates that the latch bolt is in the strike *and* that the solenoid plunger is blocking the strike. The yellow light indicates the latch bolt is in the strike but is unblocked, allowing access. The red light signals that the latch bolt is out of the strike.

You have seen several of the more common among a dozen-plus types of electrical strike release units. Let's consider that you have installed one or more of these types, and that for one reason or another, you have some trouble—after installation—with the unit. What do you do?

Troubleshooting Adams–Rite Electric Strike Latch Units

Accurately checking an electrical circuit, just as any other job, requires specific tools to perform. Make sure you have a *good* 20,000 volts per ohm volt-ohm-milliammeter. The Simpson model #261, one of several

available on today's market, will run you about $60, but it is money well spent, especially when you will be installing and maintaining electric strike latch units.

First, read the volt-ohm-milliammeter (vom) instructions and make some practice runs on simple low voltage circuits to ensure that you properly understand the operating instructions and can correctly read the results on the meter.

1. *How to check voltage.*
   a. Zero the pointer.
   b. Be sure the power is turned off to the circuit being measured.
   c. Set the function switch to the correct voltage to be measured (+dc or ac).
   d. Plug the black test lead into the common (−) jack; plug the red test lead into the (+) jack.
   e. Set the range selector to the proper voltage scale. *Caution:* It is very important that the selector be positioned to the nearest scale voltage to be measured.
   f. Connect the black test lead to the negative side

of the circuit and the red lead to the positive side. This is applicable to the dc circuit only. Turn the power on to the circuit being tested. If the pointer on the vom moves to the left, the polarity is wrong. Turn the switch function to −dc and turn the power back on. The pointer should now swing to the right for the proper reading on the dc scale. For an ac circuit, the connections are the same except as noted in c above. You don't have to worry about polarity in the ac circuit.

g. Be sure to turn the circuit off before disconnecting the vom.

2. *How to check current (dc)*.

a. Zero the pointer.

b. Be sure the power is turned off to the circuit being tested.

c. Connect the black test lead to the −10A jack and the red test lead to the +10A jack.

d. Set the range selector to 10 amps.

e. Open the circuit to be measured by disconnecting the wire that goes to one side of the solenoid. Connect the meter in series—that is, hook the black lead to one of the disconnected wires and the red lead to the other wire.

f. Turn the power on to the circuit and observe the meter. If the pointer moves to the left, reverse the leads in the jacks.

g. Turn the power back on to the circuit and read amperage on the dc scale.

h. Turn the vom off before disconnecting.

3. *Checking for line drop*. Measure the line drop by comparing voltage readings at the source (the transformer's secondary or input side) with the reading at the electric strike connection.

4. *How to find circuit shorts in a hurry*. Again the vom is the most reliable instrument for detecting a circuit short. This is accomplished by setting up the vom to measure resistance.

a. Set the range switch at position R×1.

Fig. 19-5. Installation details for a key-in-knob latch unit. (courtesy Adams Rite Mfg. Co.)

JAMB PREPARATION

Fig. 19-6. Another wooden door jamb preparation. (courtesy Adams Rite Mfg. Co.)

b. Set the function switch at +dc.

c. Connect the black test lead to the common (−) jack and the red lead to the + jack.

d. Zero the pointer by shorting the test leads together.

e. Connect the other ends of the test leads across the resistance to be measured—in the case of a solenoid, one of the test leads to one coil terminal, and the other test lead to the other terminal.

f. Watch the meter. If there is no movement of the pointer, the resistance being measured is open. If the pointer moves to the peg on the right hand side of the scale, the resistance being measured is shorted closed. If you get a reading in between these two extremes, it is very likely the solenoid is okay.

What to do when:

1. The strike will not activate after installation:

a. Check the fuse or circuit breaker supplying power to the system.

b. Check to make sure that all wiring connections are securely made. When wire nuts are used, ensure that both wires are firmly twisted together in order for good electrical transfer between the wires

c. Check the solenoid coil rated voltage (as shown on the coil label) to make sure that it corresponds to the output side of the transformer within ±10%.

d. Using the vom, check the voltage at the secondary (output) side of the transformer.

e. Using the vom, check the voltage at the solenoid. This would assure you that there are no

291

Fig. 19-7. Strike unit dimensions for wood door jamb installations. (courtesy Adams Rite Mfg. Co.)

broken wires, bad rectifiers, or bad connections.

    f. Check the coil for a possible short.

2. Transformer overheats:

    a. Make sure that the rated voltage of the transformer and the rated voltage of the coil correspond within ±10 percent.

    b. Make sure the VA rating is adequate. For all Adams-Rite units, it is recommended you use 40 VA. 20 VA is the absolute minimum and you could experience transformer heating in even moderate use applications.

3. Rectifier overheats:

    a. The rectifier is wired wrong. This means the overheating is of a temporary nature (a few milliseconds) and then it's burned out; or,

Fig. 19-8. Typical wiring diagram for an audible (ac) operation. (courtesy Adams Rite Mfg. Co.)

Fig. 19-9. Same circuitry, but with a rectifier added for silent (dc) operation. (courtesy Adams Rite Mfg. Co.)

In this, one of many possible monitoring arrangements: The Green light indicates that latchbolt is in strike solenoid plunge is blocking strike. Yellow light indicates latchbolt is in strike but strike is unblocked, allowing access. Red light signals that latch bolt is out of strike.

Fig. 19-10. Wiring for a monitor signal operation. (courtesy Adams Rite Mfg. Co.)

b. There are too many solenoids being supplied by a single rectifier and you are pulling more current through it than the system diodes are rated for.

4. Solenoid overheats:

a. The coils used in an electric strike latch, when used as rated (continuously or intermittently by pulsing once per second) gave a coil temperature rise rating of 65°C (149°F) above ambient. That means 149°F plus 72°F, for example, equals 221°F. The coil insulation is rated at 130°C, which is 266°F. Regardless of whether the coil goes to 221° or 266°, we're talking about something too hot to keep your fingers on!

b. With the above in mind, I think it is fair to say the vast majority of intermittent duty units never see that kind of use. If you have a coil that gets extremely hot on very short pulses at two or three second intervals, you either have the wrong coil or the wrong transformer output. The same is basically true for continuous duty coils. If the coil temperature exceeds the ratings, it has to be because the coil voltage or the transformer are improperly coordinated.

c. If you set the meter up as if testing for a short (see above) and obtain the exact resistance, you can compare to the specifications found in Fig. 19-11; this becomes another way of knowing

you have the correct coil. Fig. 19-11 also provides all the power data for the various electric strikes from Adams Rite that you might ever be required to install on various premises.

The last topic to cover concerns the correct determination of the proper wire size to use for electric strikes. Some people—especially those not trained in the various phases of proper electricity—assume that a low voltage means you can use the smallest and cheapest wire size available. This is certainly not the case. The size of the wire is dependent upon the use, the amount (voltage) of electricity going through the wire, the distance of the wire run, and also the number of strikes to be used on the wire run.

To determine the correct wire size, obtain the wire size factor from table 1 of Fig. 19-12 and multiply this number by the distance to the strike (in feet). Take the resulting number in circular mills (cm) to determine the wire size required from table 2 of the figure.

The following two examples will show you what I mean:

Example 1: A 24 volt ac intermittant (24 Vac-int) is to be mounted 60 feet from the transformer. The wire size

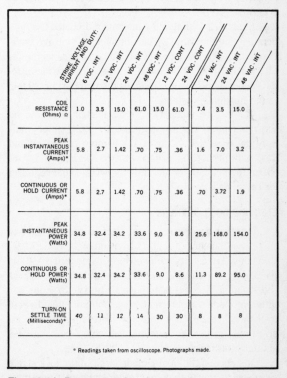

STRIKE VOLTAGE CURRENT AND DUTY:	6 VDC - INT	12 VDC - INT	24 VDC - INT	48 VDC - INT	12 VDC - CONT	24 VDC - CONT	16 VAC - INT	24 VAC - INT	48 VAC - INT
COIL RESISTANCE (Ohms) Ω	1.0	3.5	15.0	61.0	15.0	61.0	7.4	3.5	15.0
PEAK INSTANTANEOUS CURRENT (Amps)*	5.8	2.7	1.42	.70	.75	.36	1.6	7.0	3.2
CONTINUOUS OR HOLD CURRENT (Amps)*	5.8	2.7	1.42	.70	.75	.36	.70	3.72	1.9
PEAK INSTANTANEOUS POWER (Watts)	34.8	32.4	34.2	33.6	9.0	8.6	25.6	168.0	154.0
CONTINUOUS OR HOLD POWER (Watts)	34.8	32.4	34.2	33.6	9.0	8.6	11.3	89.2	95.0
TURN-ON SETTLE TIME (Milliseconds)*	40	11	12	14	30	30	8	8	8

* Readings taken from oscilloscope. Photographs made.

Fig. 19-11. Power data for electric strike units. (courtesy Adams Rite Mfg. Co.)

STRIKE USED	WIRE SIZE FACTOR		CM	WIRE GAUGE
6 VDC — INT	128		1000 AND BELOW	20
12 VDC — INT	50		1001 — 1600	18
24 VDC — INT	13		1601 — 2550	16
48 VDC — INT	3.5		2551 — 4100	14
12 VDC — CONT	14		4101 — 6500	12
24 VDC — CONT	3.5			
16 VAC — INT	10			
24 VAC — INT	34			
48 VAC — INT	9			

Fig. 19-12. Tables provide data for determining wire size required based on type of electric strike unit used. (courtesy Adams Rite Mfg. Co.)

factor for the unit is 34. Multiply 34 times 60 and this equals 2040. 2040 on table 2 corresponds to a wire size of 16 gauge.

Example 2: What is the maximum distance that you could mount the same 24 Vac int using a 20 gauge wire? From table two, the largest CM number corresponding to 20 gauge is 1000. This, divided by the wire size factor for 24 Vac-int, which is 34, produces the number 29; 29 feet is the maximum distance.

In the case when two or more strikes are to be actuated simultaneously on the same circuit, simply multiply the wire size factor by the number of strikes and then multiply by the distance to the farthest strike to obtain the cm number and thus the wire gauge required for the installation.

ELECTRIC DOOR OPENERS

Another major supplier of electric door openers in the Trine Consumer Products Division of the Square "D" Company. Trine units are available for standard wood and metal doors, in addition to the narrow stile doors, such as those obtainable from Adams-Rite (Fig. 19-13). Fourteen different models are currently available for such diverse application as entrance doors to apartments, banks, institutions, industrial buildings and complexes, offices, and interlocking door control.

Installed in the door jamb in place of the lock strike or in conjunction with various other locksets, these units offer the ability to lock or unlock a door from a remote control station. This variety of electric strikes cna be used with most standard brands of locksets in both mortise and surface mount types.

Remember that electric strike units have two systems of operation. For unlocking doors (normally in the locked position) it opens when energized. For locking doors (normally unlocked) the door is locked when the unit is energized.

For unlocking doors, the units come in standard voltages: 3 to 6 Vdc and 8 to 16 Vac. These units are electromagnetic coil types. For locking doors they come in standard voltage of 24 Vac, with a 47 ohm solenoid, and 24 Vac 119 ohm solenoid. These units are suitable for constant duty operation without overheating or burnout.

Both types of units are fail-safe in that they will automatically go to the unlock position in the event of power failure, which assures an unlocked passageway in case of emergency, such as a fire on the premises.

Figure 19-14 provides the electrical operating characteristics for both the electromagnet and solenoid types of units. Figures 19-15 through 19-17 illustrate several models that are commonly installed by locksmiths in the course of work.

At Fig. 19-18 and 19-19 are specification guide details relating to the actual full-size opening of the various types of Trine electric door opener latch cavities. Remember that when you install one, check the lock latch width and throw to the cavity size of the door opener selected, allowing for a minimum 1/32″ clearance at all

Fig. 19-13. Electric door openers come in a variety of sizes and shapes to fit different installation application needs. (courtesy Trine Consumer Products Division, Square "D" Company)

points. For lock latch height for the unit installed, always be sure to check the detailed drawing provided with the lock latch to ensure that you have properly cut the cavity and not over-or under-cut. For the lock latch heights, see the detail drawings in Fig. 19-20 and 19-21. A typical installation detail, for example, using the Trine Model 001, is shown at Fig. 19-22. Jamb preparation for the installation of an electric strike in a metal door frame is at Fig. 19-23.

During jamb preparation, a neat cutting job is required for good appearance, so special care must be taken to ensure that the dimension of the height is adequate for a loose fit insertion in the cutout. Too tight a fit here will cause the binding of the keeper in the case and result in poor performance.

The dimension indicated by an asterisk must be determined by the installer or the jamb fabricator; it is dependent on the lockset used. This dimension estab-

ELECTRICAL OPERATING CHARACTERISTICS
Average Current Ratings

ELECTRO-MAGNET TYPE
(normally LOCKED . . . UNLOCKED when energized)

AC Voltage	*Coil-Ohms Resistance	Current Draw	DC Voltage	*Coil-Ohms Resistance	Current Draw
10	2.0	1.95	6	2.0	3.0
16	2.0	2.0			
**24	15.0	.66	**24	15.0	2.0
48	73.0	.18	48	73.0	.66

* Plus or minus 10%.
** Suitable for CONSTANT DUTY OPERATION.

SOLENOID TYPE
Available (normally LOCKED) and (normally UNLOCKED)

AC Voltage	*Coil-Ohms Resistance	Current Draw	DC Voltage	*Coil-Ohms Resistance	Current Draw
FOR INTERMITTENT USE					
24	47.0	.42	24	119.0	.26
			48	200.0	.10
FOR CONSTANT DUTY OPERATION					
24	119.0	.26	24	119.0	.26

*Plus or minus 10%

Fig. 19-14. Electrical operating characteristics for Trine units. (courtesy Trine Consumer Products Division, Square "D" Company.)

*No. 014
For ANSI - Key In-Knob Lock Sets
Std. 115.3
2-7/8″ high
*Patent applied for

Fig. 19-15. Trine model 014 for key-in-knob locksets. (courtesy Trine Consumer Products Division, Square "D" Company)

lishes the relationship between the keeper and the bolt. Incorrect location of the strike will prevent the door from shutting tightly or cause excessive binding between the keeper and the bolt. Leave approximately 1/32″ free space between the keeper and the bolt.

While it may seem simple and easy to do (which it is), the two strike wires must be attached securely by wires nuts to the supplied voltage lines. Be careful not to damage the wires when inserting the strike in the jamb cutout. This can be accomplished by hooking the wire outlet side under the jamb edge first and then pushing the other side of the strike in proper position.

Low Voltage Electric Door Opener Circuitry

Figure 19-24 illustrates and provides specifics for audible (ac) operation, while Fig. 19-25 is for silent (dc) operation.

While it seems impracticable, some businesses even use the automatic electric interlock type of circuitry for rooms that serve more than one consumer, such as a bathroom between two offices, or separate businesses

located in the same building. Figure 19-26 shows how the electric interlock would work in such a situation. Such interlocks can also be used for air shower rooms, money counting rooms, x-ray and photographic rooms, radiation and biological labs, and other such type installations.

In such a case as shown in Fig. 19-26 the door opener is normally unlocked, but becomes locked when energized. The door indicator lights (DL-1/DL-2) can be used

Fig. 19-16. Solenoid coil unit for high-intensity use. (courtesy Trine Consumer Products Division, Square "D" Company)

Fig. 19-17. Model 003 with a magnet coil for normal use; this unit is used with rim type locksets. (courtesy Trine Consumer Products Division, Square "D" Company)

if visual signal is required. The opening of either door operates the door switch (a mortise type, Trine #340 or surface type, Trine model #316). It also activates the opposite unlocked door opener, locking the door. When the door shuts, the opposite door returns to the unlocked position.

Returning now to the use of the unit in a communicating bathroom locking situation, upon entering the bathroom the occupant turns the outer knob, unlocking the lockset. To lock the bathroom's doors, the occupant closes both doors if open, and operates the wall switch to locked position, illuminating both DL-2 door indicator lights and locking both doors. Operating the wall switch to the unlocked position turns off both door indicator lights and unlocks both the doors. After leaving the bathroom, the occupant presses the pushbutton in the lockset, preventing anyone from the adjoining room from entering the occupant's room. An emergency release is obtained by installing a push button at each door which will unlock its own door.

While this is only one type of situation where more than one door opener (always an RS type unit in such interlocking cases) is used, there are other instances where multiple door openers are required, with each having its own pushbutton control switch; further, each one is operated off the same power line. Figure 19-27 illustrates how the wiring of multiple door openers is accomplished.

Fig. 19-18. Size specifications for various latches. (courtesy Trine Consumer Products Division, Square "D" Company)

NO. 001

No. 002/S002

Nos. 014/014W

Fig. 19-19. More latch cavity specifications. (courtesy Trine Consumer Products Division, Square "D" Company)

Electrical ratings for solenoids for fail-safe (continuous duty) and standard (intermittent duty only) units are at Fig. 19-28.

Due to the remote control applications allowed for electrical strike units, there can be a considerable distance between the strike location and the actual control for that particular strike. At Fig. 19-29 are the *minimum* wire sizes required for standard and fail-safe electrical strike unit voltage at various distances.

Troubleshooting Trine Electrical Strikes

Earlier we discussed troubleshooting procedures, but no matter how carefully units are tested and retested at the point of manufacture, problems can still arise. These troubleshooting procedures concern the unit after installation, and can relate to any electrical strike unit in use.

1. After installation, the strike hums but does not operate properly when the power is applied.

Probable cause: ac voltage has been supplied, but the rectifier has not been installed between the transformer and the strike.

Correction: Install the rectifier per the wiring diagram at Fig. 19-30.

2. After installation, the strike does not operate even if the power is left on for one minute (maximum time) and the strike is noticeably warm.

Probable cause: (1) Static loading against the door is excessive due to weather stripping, a warped door, wind loading on the door, or improper jamb preparation. (2)

No. 8 F.H. Screw

No. S005
Mortise type for Schlage type (bored cylindrical) lock sets. 6" wire leads.

CONCEALED MOUNTING SCREW No. 8R.H.

No. 003
Narrow, surface type. For mounting in conjunction with rim type locksets. 6" wire leads.

Fig. 19-20. Detail specification drawings for magnet coil units. (courtesy Trine Consumer Products Division, Square "D" Company)

Fig. 19-21. Specification details for high-intensity solenoid coil electric door opener units. (courtesy Trine Consumer Products Division, Square "D" Company)

TYPICAL INSTALLATION DETAIL

No. 001
Door Opener
Door 1¾" (44)
Latch throw
½" (12.7)
Door Jamb 2" (51)

Magnet coil door openers are for normal intermittent use.

Fig. 19-22. Typical installation detail for model 001 magnet coil unit. (courtesy Trine Consumer Products Division, Square "D" Company)

115 VOLT AC LINE

TRANSFORMER

Specifier should select the door opener to suit the door frame and the door lock set. Electrician is to provide the power at the point of installation. The buzzing sound in ac operation is usually not considered offensive in intermittent use. It acts as an "Open" signal.

PUSH BUTTON

DOOR OPENER

Fig. 19-24. Low voltage electrical circuit for an audible operation. (courtesy Trine Consumer Products Division, Square "D" Company)

Insufficient voltage is reaching the strike due to an inadequate supply wire size.

Correction: (1a) Adjust the strike bolt relationship so that some slight gap exists in the locked position. (1b) Balance air conditioning pressures on opposite sides of

the door. (2) Provide sufficient sized service wire (Fig. 19-29).

3. The keeper moves to fully open on power application, but does not return to the locked position when the door closes.

1¼"
⅝"
4⅞"
4⅛"
3⅝"
3-7/16"
12-24 TAPPED HOLES (2) *
1-11/16"

Fig. 19-23. Typical procedural details for a mounted door frame installation. (courtesy Trine Consumer Products Division, Square "D" Company)

115 VOLT AC LINE

TRANSFORMER

RED LEADS

SR-1

SR-1 rectifier units are recommended when a continuous duty system is specified. Door indicator lights DL-1/DL-2 can also be used if a visual signal is required. SR-1 rectifiers silence the door opener . . . only a "click" is heard when energized. SR-1 can be mounted at secondary terminals of transformer or wired to door opener leads. Up to 3 door openers can be operated with each SR-1 rectifier.

PUSH BUTTON

SR-1

RED LEADS

DOOR OPENER

Fig. 19-25. Low-voltage circuitry for silent operation of unit. (courtesy Trine Consumer Products Division, Square "D" Company)

Fig. 19-26. Automatic electric interlock for the operation of opposite side doors in an area serving more than one user. (courtesy Trine Consumer Products Division, Square "D" Company)

Probable cause: If the deadbolt is extended, but does not pull the keeper to the fully locked position on door closure, the return spring may need adjusting.

Correction: Deform the return spring slightly by forcing it toward the keeper.

4. The keeper binds in the strike case on both opening and closing.

Probable cause: The cutout in the jamb is too narrow and is forcing the strike case walls inward toward the keeper.

Correction: Increase the vertical dimension until the keeper operates freely.

It should be noted that electric door openers are also available and work extremely well for deadbolts. They can be used with just about any manufacturer's lockset. They include the following types:

☐ Mortise locks with deadbolt.
☐ Mortise lock with deadlatch and garded latch.
☐ Individual deadbolt lock.
☐ Panic exit devices.
☐ Maximum security deadbolts (¾" throw).
☐ Release up to 1" throw deadbolts.
☐ With optional adapters, they may be used with cylindrical locks, alarm locks or with Detex bolts.

These units will also allow for the strike operation of mortise locks, with both latchbolt and deadbolt release by the action of a single keeper. The keeper is held open after the door is opened if the deadbolt was in an extended position at the time of the door opening. The keeper automatically returns to the locked condition when the deadbolt reenters the strike. The door may also be closed with deadbolts extended. In strikes with options H, J and K (Fig. 19-31), the keeper is forced open by the extending latchbolt and springs closed to the locked position to accommodate the returning latchbolt. Other options are also shown at Fig. 19-31.

Figure 19-32 illustrates two Trine unit models (015 and 016) specifications. Along with the basic unit for installation, the necessity of a heavy duty transformer (Fig. 19-33) is necessary to convert the primary voltage to the operating voltage.

ELECTRIFIED MORTISE LOCKS

The electrified mortise lock provides for a wider number of additional installation sales and servicing possibilities, especially when access security controls for your customer are concerned. The electrified mortise

Fig. 19-27. Circuitry for multiple wiring of door openers. (courtesy Trine Consumer Products Division, Square "D" Company)

ELECTRICAL RATINGS FOR SOLENOIDS	FAIL SAFE UNITS (CONT. DUTY)		STANDARD UNITS* (INTERMITTENT DUTY ONLY)		
	24 VDC	12 VDC	24 VDC	12 VDC	48 VDC
RESISTANCE IN OHMS	115	33	16	4	90
WATTS SEATED	4.8	4.4	36	36	25
AMPS SEATED	.2	.37	1.5	3.0	.53
	*Bridge rectifier may be supplied with any unit for ac operation.				

Fig. 19-28. Electrical ratings for intermittent use and constant duty operation for solenoid coil door openers. (courtesy Trine Consumer Products Division, Square "D" Company)

lock (Fig. 19-34) is another excellent and very durable product that has minimal installation procedures, yet offers long life and reliability for your customer.

This lock unit has the standard rugged lock case construction of stainless steel plus a normal mortise strike. The inside knob is always free, while the outside knob is rigid except when the 24 Vdc internal solenoid is energized, which releases the outside knob, allowing for latchbolt retraction. Like any other mortise lock which is standard, it is reversible, allowing for different hands.

Installation of the Electrified Mortise Lock:

1. Mark the door and jamb (Fig. 19-35). Mark the horizontal centerline of the lock on both sides and the edge of the door. Mark the vertical centerline of the lock

at the door edge and the vertical centerline on both sides of the door as measured from the door edge.

2. Mortise the door for the lock case and front (Fig. 19-36). Refer to the installation template that comes with the lock unit for the specific dimensions.

3. As indicated on the template and as shown in Fig. 19-36, drill the required holes. *Caution:* Check the lock for function, hand, and bevel before drilling. Mark and drill only those holes required for your function.

4. Remove the front plate and mount the lock in a mortised opening. Apply the trim (four trim variations are possible; see Figs. 19-38 through 19-41 at the end of the installation instructions). Apply the cylinder (using a minimum of four turns), then fasten with the cylinder clamp screw. Replace the front plate. *Caution:* Check the cylinder and lock for proper function before closing the door.

5. Install the strike (Fig. 19-37). Position the strike on the door jamb as shown. Mortise the jamb and apply the strike. When a box strike is used, mortise deep enough to allow the latchbolt and deadbolt to fully extend.

Reversing the Lock and Bevel:

At times you may be required to replace or remove and reinstall the lock on another door.

Fig. 19-29. Electrical strike voltage (fail-secure mode of operation). (courtesy Trine Consumer Products Division, Square "D" company.)

Fig. 19-30. Circuitry installation of a rectifier to alleviate a strike hum. (courtesy Trine Consumer Products Division, Square "D" Company)

"G" LOCK ADAPTER KIT

Deadlocking Lever

OPTION "G"

Spring clip with two mounting screws attaches to inside of strike case at either end.

DEADBOLT BARRIER KIT

OPTION "H"

Adapter plate with four mounting screws and reverse wound bias spring. Provides HOTEL LOCK-OUT function. Latchbolt is released but deadbolt is not. Keeper is spring biased to locked position.

NOTE: LOCKSET AND HANDING MUST BE SPECIFIED

ALARM LOCK OR DETEX APPLICATION

NO. 016

Door Swing

Door

No. 016   Alarm Lock

Mounting Adapter Block

Fire Safety with Access Control

DEADLOCKING LEVER KIT

OPTION "K"

Adapter plate with two mounting screws and reverse wound bias spring. Fills in strike opening to insure depression of auxiliary dead locking lever. Keeper is spring biased to locked position.

CYLINDRICAL LOCK KIT

OPTION "J"

Adapter plate with four mounting screws and reverse wound bias spring closes down strike opening so that only latchbolt is accommodated. Keeper spring is biased to locked position.

Fig. 19-31. Various strike options for electric door openers used with deadbolt locks. (courtesy Trine Consumer Products Division, Square "D" Company)

No. 015

No. 016

3/32"   1⅝"
2 1/16"

4⅞"   4⅞"

3⅜"
KEEPER OPENING
3¼"

No. 12 Flat Head Screw

1⅝"

3/32"   1⅝"
2 1/16"

3⅜"
KEEPER OPENING
3¼"

1⅞"

DOOR

Typical Installation

1/32"

DOOR

Typical Installation

1/32"

Fig. 19-32. Technical specifications for two Trine door opener models that can be used with any manufacturer's lockset, (courtesy Trine Consumer Products Division, Square "D" Company)

Fig. 19-33. Heavy-duty transformer for voltage conversion. (courtesy Trine Consumer Products Division, Square "D" Company)

1. Remove the cover from the lock.

2. Reverse the latchbolt and the auxiliary bolt.

3. Reverse the hubs.

4. To change the bevel, loosen the two screws on the top and bottom of the case. Adjust the proper bevel and tighten the screws.

5. Replace the cover. Be certain that all parts are in their proper place before fastening the exterior screws. *Caution*: When the hand of the lock is changed, make sure that the proper hand strike is also used. A left-hand lock requires a left-hand strike; a right-hand lock requires a right-hand strike.

Figures 19-38 through 19-41 show the various trims available for this unit. Figure 19-38, the heavy duty screwless trim; Fig. 19-39, the lever handle trim; Fig. 19-40 is the medium duty trim with a Glenwood escutcheon; and Fig. 19-41 is the lever handle trim with a Utica escutcheon.

Fig. 19-34. Electrified knob lock. (courtesy Alarm Lock Corporation)

Fig. 19-35. Marking the door for installation. (courtesy Alarm Lock Corporation)

ELECTRIFIED KNOB LOCK

The electrified knob lock (Fig. 19-42) by the Alarm Lock Corp. is a heavy-duty lockset with a key bypass in the outside knob. It has a deadlocking latchbolt and stainless steel trim for long-lasting exterior or interior use against the elements. The inside knob is always free for exit. The outside knob is rigid to prevent entry except when the internal solenoid is energized, which releases the outside knob to retract the latchbolt and open the door. The electrified knob lock has a continuous duty solenoid that uses 24 Vac power supply (360 milliamps).

Installation Procedures:

1. Mark the door. Mark the height line on the edge of the door 38″ from the floor (or other height that may be required by the customer; this is the suggested height, although it may vary according to the individual needs). Position the centerline of the template that comes with

Fig. 19-36. Bore and mortise for latch front. (courtesy Alarm Lock Corporation)

Fig. 19-37. Strike installation details. (courtesy Alarm Lock Corporation)

the lock on the height line and mark the center point of the door thickness and the center point for the 2⅛″ hole through the door (Fig. 19-43).

2. In boring the holes, bore the 2⅛″ hole at the point marked from both sides of the door. Bore the 1″ latch unit hole straight into the edge of the door from the center point of the height line. Mortise the latch front and install the latch unit (Fig. 19-44).

3. Removing the lock from the box, remove the inside trim. This is accomplished by inserting the small end of the spanner wrench through the hole in the knob bearing sleeve. This hole faces the latch retractor. Press the knob catch to slide the knob off the spindle and remove the rose from the threaded sleeve (Fig. 19-45).

4. The door thickness adjustment is accomplished by adjusting the outside rose either in or out until the indicator on the lock shows the proper thickness for the door. Make sure the rose clicks into place (Fig. 19-46). The lock will fit on all doors 1⅜″ to 2″ thick. (If the lock you require is for a door of greater thickness, it will have to be obtained as a special request to the company.)

5. Next, install the main unit Fig. 19-47). The main housing must engage with the latch prongs and retractor with the latch tailpiece as shown in Fig. 19-48.

6. Secure the trim by replacing the rose and screwing it on to the threaded spindle with your fingers as far as it will go. Then tighten the rose with the spanner wrench (Fig. 19-49).

7. Replace the knob by lining up the keyway in the knob with the slot in the spindle. Then push the knob

NO.	PART NAME	NO REQ'D
1	OUTSIDE KNOB AND SPINDLE ASS'Y	1
2	NYLON BUSHING	2
3	ROSE BUSHING	2
4	ROSE	2
5	OUTSIDE MOUNTING PLATE ASS'Y	●
6	INSIDE PLATE ASS'Y	1
7	MOUNTING SCREWS	2
8	INSIDE KNOB ASS'Y	1
9	SLEEVE & PIN ASS'Y	1
10	INSIDE KNOB COLLAR	1
11	WRENCH	1

Fig. 19-38. Heavy-duty screwless trim. (courtesy Alarm Lock Corporation)

straight on until the knob hits the knob catch. Depress the knob catch and push the knob straight on until the knob catch clicks into place. Test both knobs to make sure they are securely fastened to the spindle (Fig. 19-50).

8. Install the strike by making a shallow mortise in the door jamb to align with the latch face and install the strike as in Fig. 19-51.

If you must reverse the cylinder of the lock, the following applies:

1. Insert the key into the cylinder, turn the key clockwise as far as it will go, and at the same time, depress the knob catch retainer. When the retainer moves down, pull the knob off.

2. Turn the knob 180° and replace it as follows: Remove the key and push the knob on the spindle as far as it will go. Insert the key and turn it clockwise to its most extreme position while depressing the knob catch retainer. When the retainer moves down, push the knob onto the spindle allowing the retainer to click into position.

If you must remove the cylinder, push the cylinder tailpiece sideways with your finger until the tailpiece unsnaps from the plug. Then turn the knob over and hit the shank of the knob with a sharp blow on a flat surface. This will release the knob ferrule and will permit the removal of the cylinder. Rekeying of the cylinder is conventional.

Now that you understand the specifics of the lock installation itself, consider the electrical considerations related to the lock. The electrical power for the lock can either be concealed within the door by use of a current transfer unit (Fig. 19-52) or left exposed on the interior side. The use of a current transfer unit is a unique means of bringing the electrical current from the lock side of the door to the door. The face and strike must be thoroughly matched and the door closed for power to flow through. With this in place, from a remote switch activation, power will flow through allowing the door to be opened.

The other method, leaving the wiring exposed, is not really what it seems. The use of an armored door loop is used and it provides a much easier means for bringing electrical current from the hinge side of the door frame to the lock in the door. It consists of an 18″ flexible armored cable which is attached to the frame on one end, and just below the lock on the inside of the door. The lock wires can be passed through the door wall via a small hole and connected to the loop wire ends.

ELECTROMAGNETIC LOCKS

While electric locks are available and used for many applications, there is also the need for electromagnetic

NO.	PART NAME	NO REQ'D
1	OUTSIDE HANDLE AND SPINDLE ASS'Y	1
2	NYLON BUSHING	2
3	ROSE	2
4	MOUNTING SCREWS	6
5	MOUNTING PLATE	2
6	INSIDE HANDLE	1
7	SET SCREW	1
8	WRENCH	1

INSIDE   OUTSIDE

Fig. 19-39. Lever handle trim. (courtesy Alarm Lock Corporation)

NO.	PART NAME	NO REQ'D
1	OUTSIDE KNOB AND SPINDLE ASS'Y	1
2	NYLON BUSHING	2
3	OUTSIDE ESCUT.	1
4	INSIDE ESCUT.	1
5	MOUNTING SCREWS	8
6	INSIDE KNOB	1
7	SET SCREWS	2
8	CYLINDER RING	1
9	CYLINDER SPRING	1
10	MORTISE CYLINDER	1
11	KEYS	2

Fig. 19-40. Medium-duty trim with the Glenwood escutcheon. (courtesy Alarm Lock Corporation)

locks. Such locks can be used in hospitals, convalescent homes, banks, universities, museums, libraries, factories, laboratories, penal institutions, and airports, among others.

The electromagnetic lock, while operating at less than 12 Vdc, can exert a holding force of 1500 pounds. This is fantastic for a lock that consists of two components—the lock itself, and its armature.

Imagine a lock that mounts rigidly to a door frame and an armature that mounts to the door (Fig. 19-53). When the door is closed, the two make contact; upon activation of the lock, the two are attracted to each other. The door, for all practical purposes, is bonded solidly to the frame.

Ten years ago, such a lock would have been priced out of reach, but today it is possible for almost every business or institution that requires such a locking device to afford one. As a locksmith, you must be prepared and familiar with this "unconventional" (in the traditional sense) lock.

Security Engineering, the leader and major proponent of this lock, has indicated the major features as follows:

☐ A high holding force (1500 pounds)—effectively seals the door from both sides, keeping intruders out and/or valuables safely in.

☐ Conserves energy—the efficient low power consumption design (2 amps) ensures a long life and economical operating cost for the lock.

☐ High reliability—recommended for high trafficked openings with extreme usage, where conventional locks, locks with moving parts, or electromagnetic locks with a lesser holding force may not be capable of withstanding the abuse.

☐ Fully fail safe—locking and unlocking is positive and instantaneous, with no residual magnetism or moving parts that might stick, jam, bind, wear out, and/or need replacement.

☐ Extremely low maintenance—no maintenance or adjustments are required after initial installation.

☐ Exclusive mounting design—this permits rapid efficient installation in new or existing buildings and automatically compensates for normal door wear, sag, warpage, or misalignment.

☐ Self-contained unit—it is completely factory wired with all options for the unit already built in, fully

NO.	PART NAME	NO REQ'D
1	OUTSIDE HANDLE AND SPINDLE ASS'Y	1
2	NYLON BUSHING	2
3	OUTSIDE ESCUT.	1
4	INSIDE ESCUT.	1
5	MOUNTING SCREWS	2
6	INSIDE HANDLE	1
7	SET SCREW	1
8	WRENCH	1
9	CYLINDER RING	1
10	CYLINDER SPRING	1
11	MORTISE CYLINDER	1
12	KEYS	2

Fig. 19-41. Lever handle trim using the popular Utica escutcheon. (courtesy Alarm Lock Corporation)

Fig. 19-42. Model #121 electrified mortise lock. (courtesy Alarm Lock Corporation)

concealed, and tamper-resistant. There is only one external wire connection to be made.

☐ Four operating voltages—12 Vdc; standarad 12 Vac; 24 Vdc; and 24 Vac.

Figure 19-54 illustrates a typical installation and the various features (standard and optional) that can come with a unit.

Series 3900 Electromagnetic Lock Mounting:

It is important that prior to actually attempting to

Fig. 19-43. Electrified knob lock installation, Step 1.

Fig. 19-44. Electrified knob lock installation, Step 2.

install the electromagnetic lock, that you handle the equipment with care, since damaging the mating surfaces of the armature of the electromagnetic may reduce its locking efficiency. The electromagnet mounts rigidly to the door frame header. The armature mounts to the door with special hardware provided that allows it to pivot about its center to compensate for door wear and misalignment. All template use must take place with the door in its normal closed position.

The lock face is covered with a rust inhibiting film. This should not be removed. If the film is accidentally removed, you should clean the lock face with a clean dry soft cloth (do not touch the lock face with your hands) and reapply a rust inhibitor such as M1 (manufactured by Starret) or LPS3 (manufactured by LPS Laboratories); these rust inhibitors are readily available in most hardware stores. Then apply a light film of rust inhibitor generously to the armature face.

1. Note the type of door frame header and install the filler place or angle bracket as required (Fig. 19-55).

2. Fold the template (Fig. 19-56) where indicated to form a 90° angle. For a swinging door, place the

Fig. 19-45. Electrified knob lock installation, Step 3.

Fig. 19-46. Electrified knob lock installation, Step 4.

BEFORE INSTALLATION MAKE SURE THAT CLEARANCE FOR ANTIFRICTION LATCH IS CUT-OUT

Fig. 19-47. Electrified knob lock installation, Step 5.

template against the door header and door opposite hinge side of the door jamb. For a pair of swinging doors, place the template against the door and door header at the center of the door opening. Transfer the hole locations to the door and door frame header.

3. Follow the template instructions for the specific hole sizes required. See Fig. 19-56 to determine the proper armature mounting hole preparation. The hole is designated as the mark ⊕ in Fig. 19-56.

4. Mount the armature to the door with the hardware provided as shown in the figures.

5. Install the mounting plate to the header with the five #10 screws.

6. Fasten the model 3900 lock to the mounting

MAIN HOUSING

LATCH TAILPIECE

RETRACTOR ENGAGES LATCH TAILPIECE

LATCH PRONGS ENGAGE MAIN HOUSING

Fig. 19-48. Electrified knob lock installation, Step 5.

TURN CLOCKWISE AND TIGHTEN FIRMLY

Fig. 19-49. Electrified knob lock installation, Step 6.

Fig. 19-50. Electrified knob lock installation, Step 7.

Fig. 19-52. Current transfer unit. (courtesy Alarm Lock Corporation)

plate with the two socket screws and compensate for any misalignment by adding or subtracting washers at the armature mounting screws.

7. Finally, firmly tighten all the screws. Install the anti-tamper plugs into the holes over each socket head mounting screw. Use a soft hammer to avoid damaging the lock case.

Wiring Instructions:

In some instances you may be using a prefabricated power supply for the lock unit. But suppose you are not. What then? To achieve the best possible electromagnet lock installation, note the following suggestions (both right and wrong ways illustrated for each procedure).

1. The proper placement of the on-off station control in the electrical circuit will ensure instantaneous door release and avoid any delay in complete unlocking (Fig. 19-57).

2. Care must be taken when adding any auxiliary controls or indicators to the electrical circuit. Improper

placement may cause a slight delay in the door release (Fig. 19-58).

3. Solid state on-off devices must be properly protected from the back voltage generated from the collapsing electromagnetic lock coil when the unit switched off. This may be accomplished by installing a metal oxide varistor (MOV) across the lock coil lead wires (Fig. 19-59). The MOV is available from Security Engineering; see the address in the Appendix B.

4. Installation of the MOV is for arc suppression in ultra high usage installations to prevent switch contact deterioration. In any situation, though, the installation of a diode across the leads is not recommended (see Fig. 19-58).

Another lock to consider is the surface-mounted

Fig. 19-53. Model 3900 electromagnetic lock. (courtesy Security Engineering, Inc.)

Fig. 19-51. Electrified knob lock installation, Step 8.

1. Model #3900 Electric Lock
2. Electric Lock Armature
3. Door Status Switch Optional*
4. Local On/Off Station Controls
5. Power Supply
6. Remote On/Off Station Controls*
7. Hazard Sensing Safety Devices*

*As System Requires

Fig. 19-54. Model 3900 installation layout. (courtesy Security Engineering, Inc.)

electrically actuated lock, the series 2200 from Security Engineering. This lock (Fig. 19-10) has been designed to offer the advantages of remote electric locking coupled with versatility limited only by the user's imagination.

For basic operation, the lock mounts to a door frame or a door and when energized, locks or unlocks the door when the solenoid-operated deadbolt projects from the unit and makes contact with a strike.

Standard features include:

☐ Fail-safe operation—if the power is interrupted, the unit automatically unlocks.
☐ Fail-secure operation—if the power is interrupted, the unit automatically unlocks.
☐ Easy installation—the unit is not handed and thus, can be mounted on any plane.
☐ Versatility—it can be used with inswinging, outswinging, sliding, overhead or rollup doors.
☐ Maintenance free—unit parts are ruggedly built and self-lubricating.
☐ Current draw—.5 amps at 24 volts.
☐ Low voltage—24 Vac standard; 24 Vdc, 12 Vac, and 12 Vdc available.
☐ Long and silent operation—the unit has a continuous duty solenoid that will function for millions of cycles.
☐ Compact size—2″×2″×6¾″.

The unit has optional features, such as a choice of one or two signal lights; or a built-in bolt position switch to indicate whether or not the bolt is extended or retracted for interfacing with automatic door equipment or signaling the lock status to a remote location unit. *Note:* The fail secure locks are not recommended where life

Fig. 19-55. Door frame header determination tells you if a filler plate, angle bracket, or nothing will be required. (courtesy Security Engineering, Inc.)

Fig. 19-56. Model 3900 lock template. (courtesy Security Engineering, Inc.)

Fig. 19-57. Model 3900 wiring, Step 1.

Fig. 19-58. Model 3900 wiring, Step 2.

Fig. 19-59. Model 3900 wiring, Step 3.

Fig. 19-60. Model 220 series surface-mounted lock. (courtesy Security Engineering, Inc.)

safety may be compromised and is also not recommended where panic crash bar hardware is the only means of emergency egress.

Lock Mounting Instructions (refer to Fig. 19-61).

1. Remove the four screws from the unit; remove the housing cover.

2. Manually extend the solenoid bolt by pushing the BPS pin forward (if equipped) or inserting a wire into the rear of the solenoid.

3. Engage the extended solenoid bolt and strike and locate as shown in the figure on the door and door frame.

4. Using the mounting holes in the lock base and strike as templates, mount the solenoid lock and strike with the screws provided. (Several strike types are available and can be used, depending upon the situation required. See Fig. 19-62.)

5. Check for the proper alignment and see that the lock is free, having no binding in operation.

6. Connect the electrical wiring per the system option or requirements.

7. Check the operation of the lock for any final adjustments; if any, adjust as required.

8. Reinstall the housing cover and secure with screws.

A more compact but just as effective lock is the 1300 series mortise solenoid lock (Fig. 19-63). This mortise solenoid activated deadbolt is designed primarily for installation in new construction, with emphasis placed on tamper resistance and concealment of the unit.

As for operation, upon activation of the solenoid, the deadbolt projects ½″ from the faceplace or retracts ½″ to flush with the faceplate, depending upon the solenoid selected and used.

Among the standard features of this lock unit are:

☐ Fail safe and fail secure operation—interrupted power and the unit automatically unlocks.

Fig. 19-61. Installation procedures for #2200 series lock. (courtesy Security Engineering, Inc.)

Fig. 19-62. Strikes. (courtesy Security Engineering, Inc.)

Fig. 19-63. Series 1300 mortise solenoid lock. (courtesy Security Engineering, Inc.)

☐ Compact size—1⅜″ diameter×4½″ long.
☐ Voltage and current draw are the same as the 2200 series.

ELECTRICAL KEYLESS LOCKS

Simplex security locks have also kept pace with the electrically operated locks and latch unit. The Simplex keyless entrance control is a wall-mounted unit control for electrical door locks (Fig. 19-64) and has a built-in adjustable time delay feature not found in other currently available marketed units. The elimination of key control problems in high-risk areas, for home or business, is a distinct and great advantage. The Simplex entrance control is an easy, yet very secure installation. The unit is secured by two rugged draw bolts that pass completely through the wall. The bolts can be cut to any desired length to permit installation in walls from 4″ to 11″ thick (Fig. 19-65). The adjustable timer delay of 1/5 to 20 seconds permits locating the control at any convenient position near or by the door.

Two styles of front mounting plates are included to meet varying installation needs. One is 3¼″ wide by 9⅝″ high; the smaller is 1¾″ wide by 8⅛″. Both front mounting plates are of stainless steel. Some security features for this unit include:

Fig. 19-64. Wall-mounted electrical control unit for door lock or latch. (courtesy Simplex Security Systems, Inc.)

□ The electrical terminals and wire leads are protected against a short circuit attach by a heavy-duty metal back cover that completely encloses the entire electro-mechanical device inside the wall.

Also in demand by many businesses is the Simplex/Unican 2000-15 model (Fig. 19-66) which can be used with many leading brands of surface-mounted panic exit devices. Another popular unit is the model TM pushbutton lock (Fig. 19-67) which can be used in both business and home.

□ A special visual shield for each entrance control unit to protect against unauthorized observation of the button sequence operation.
□ If the control knob is forced by pliers or other tools, it will spin free without activating the electrical circuit.

Fig. 19-65. Lock unit can be mounted on any thickness wall up to 11″. (courtesy Simplex Security Systems, Inc.)

Fig. 19-66. Unican 2000-15 unit used with an emergency exit panic bar. (courtesy Simplex Security Systems, Inc.)

Fig. 19-67. Model TM lock installed in an office cabinet. (courtesy Simplex Security Systems, Inc.)

Fig. 19-69. Programmer unit.

Fig. 19-68. Pushbutton control unit.

Fig. 19-70. Pushbutton unit with card reader for increased security.

A sophisticated digital access control system that does not require a console board or monitoring station is available which the locksmith can easily install. The Memorilok system has two basic units, an access control keyboard (Fig. 19-68) and a program unit (Fig. 19-69). With these two units, 10,000 combinations are available to the customer. Now there is a third optional piece of equipment, the card reader (Fig. 19-70), which provides a substantially higher degree of security integrity for the area to which access is desired.

The Memorilok system incorporates a solid state electronic memory for the storage of the selected code combination. The entry code combination is electronically stored. Unlike other access control systems, there are no wheels, dials, or other visual indicators to betray the code combination.

Standard unit features include:

☐ *Automatic address*. A pre-set timer is activated when the first digit of the code combination is entered. The remainder of the code entry must be completed before the expiration of this time, or the entry will be can-celed and the programmer will activate the penalty time.

☐ *Automatic penalty*. A penalty time of approximately five seconds is incurred if an incorrect key is depressed during the code entry, or if two or more keys are pressed simultaneously. This feature defeats random button manipulation.

☐ *Code bypass*. There is provision within the access control unit for the connection of an external (remote) switch or pushbutton to operate the lock via the programmer unit without the use of the entry keyboard. This normally would be sited within the protected access area and be used for giving access to personnel or visitors who do not possess the code combination, but for whom access is warranted and authorized by the customer.

☐ *Memory Protection*. The Memorilok keeps the code combination stored and undamaged in case of a main power failure. This is due to its constantly charged nickel cadium battery that is contained within the unit.

Optional extra features include:

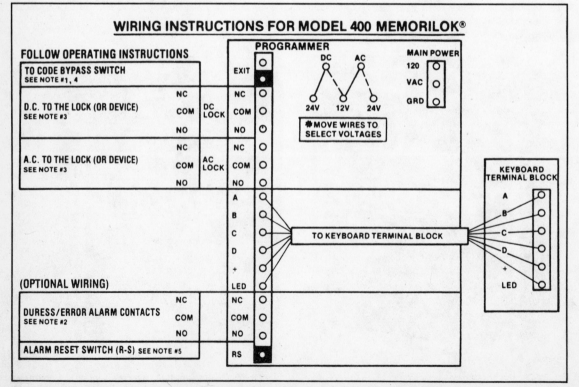

Fig. 19-71. Memorilok wiring instructions (refer to text for "notes"). (courtesy Alarm Lock Corp.)

SPECIFICATIONS:

Dimensions: Keyboard Unit	5 ¹⁄₁₆″ (128 mm) Wide 6 ¹⁄₁₆″ (153 mm) High 2 ⅝″ (66 mm) Projection
Programmer Unit	Same as Keyboard
Shipping Weight:	Complete Memorilok - 6 pounds Battery Standby Unit - 6 pounds
Main Power Input:	120 Vac or 240 Vac 50 /60 HZ. Specify voltage required.
*Output to Lock:	12 or 24 Vac or Vdc at 1 Amp. Voltage is user selectable.
Program Key Switch:	Contacts are closed in the ''ON'' position for code programming.
Wiring Information:	Includes ten feet of six conductor 22 gauge wire.

Fig. 19-72. Memorilok #400 unit specifications. (courtesy Alarm Lock Corp.)

KEY ENTRY		No. of Switch elements
	Spring Return	1 2 3 4
	One Key Entry 9h	1 2 3 4
	One Key Entry 6h	1 2 3 4
	Two Key Entry	1 2 3 4

Fig. 19-74. Four different key positions provide an opportunity for different control features. (courtesy Kaba Locks, Ltd.)

☐ *Duress/error alarm.* The duress/error alarm option is available as an integral part of the Memorilok unit, and if desired by the customer, must be ordered as original equipment. The duress alarm is activated by depressing a factory fixed last digit (from 0 to 9) instead of the last digit of the selected code combination. If entry has to be made under duress, the first three digits of the programmed code combination plus the predetermined last digit—the Duress Digit—is entered on the keyboard. This automatically operates a set of Form C (Changeover) dry contacts which, in turn, triggers any desired alarm system in effect: audible, silent, remote, or local.

The error alarm circuit will react upon any number of selected unsuccessful attempts at entry. As soon as the predetermined number of incorrect attempts (from one to eight) have been made, the alarm is activated, warning that the system is being tampered with.

☐ *Flush mounting.* To flush mount the programmer, keyboard, or even the battery standby, an oversized front panel is used. The panels, 8″×9″ are affixed to the standard front panels in order to conceal the roughout opening around the Memorilok unit.

Fig. 19-73. Micro Kaba rotary switch. (courtesy Kaba Locks, Ltd.)

Fig. 19-75. Two-pole switch. (courtesy Kaba Locks, Ltd.)

Fig. 19-76. Specifications for the two-pole switch. (courtesy Kaba Locks, Ltd.)

The optional card reader fits into the front of the keyboard panel. It provides higher levels of security integrity because the coded card *must* be used by the individual seeking entry. It must be entered into the Memorilok before the code combination will be accepted by the electronic memory circuit.

Wiring details are in Fig. 19-71. The following notes apply, as referenced in the figure:

Fig. 19-77. Surface-mounted remote control station with a waterproof key cover. (courtesy Kaba Locks, Ltd.)

1 -/2 Tamper Loop
3 - 4 12 Vdc LED "on"
6 7 8 ON/OFF Changeover Switch. KS64

WIRING DIAGRAM FOR REMOTE
CONTROL STATIONS KS-RCS/1

1 - 2 Tamper Loop
3 - 4 12 Vdc LED "on"
6 7 8 ON/OFF Changeover Switch KS64

WIRING DIAGRAM FOR REMOTE
CONTROL STATIONS KS-RCS/2

Fig. 19-78. RCS/1 and RCS/2 wiring diagrams. (courtesy Kaba Locks, Ltd.)

Fig. 19-79. The narrow remote control switch allows for many different installation applications. (courtesy Kaba Locks, Ltd.)

Fig. 19-80. Remote control station housing mounting box. (courtesy Kaba Locks, Ltd.)

1. These connections are not used unless the remote entry/exit is required. Use the open switch.

2. All contacts are shown (NO and NC) without the lock being energized, and without being in the alarmed state.

3. On installations where there is a long wire run between the programmer and the lock or other device being operated, carefully compute the minimum re-

quirements of the lock or device and the voltage drop which will occur in the wire.

4. Code bypass switch requires a momentary switch with normally open (NO) switch contacts.

5. The alarm Reset Switch (R-S) requires a momentary switch with normal switch contacts. This should be wired between terminal #2 of the exit and terminal R-S indicated on the diagram as "blacked out

Fig. 19-81. Mounting box specification data. (courtesy Kaba Locks, Ltd.)

Rated Voltage (V)	Non-Inductive Load		Inductive Load	
	Resistive (A)	Lamp (A)	Inductive (A)	Motor (A)
ac 110 to 125	5	1.5	3	2.5
220 to 250	5	1.0	2	1.5
dc 6 to 8	5	2	5	3
110 to 125	0.4	0.05	0.4	0.05
220 to 250	0.2	0.03	0.2	0.03

Fig. 19-82. Electrical specifications for the remote control station. (courtesy Kaba Locks, Ltd.)

KEY-ACTUATED SWITCHES

The full-service locksmith should have available for customers, especially those in business, a variety of key actuated switches. Figure 19-73 shows a MicroKaba rotary switch. This miniature switch is operated by the 12-mm diameter cylinder, which has over 10,000 different possible key combinations. Four key entry variations are available, each having up to four switch elements (Fig. 19-74).

Two-pole switches (Fig. 19-75) are also available, which provide for 90 degree movement with two key withdrawal positions, thus allowing the electronic unit to be held in either the on or off position. The unit can take up to 250 volts ac at 4 amps. Dimensions of the unit are at Fig. 19-76.

The surface-mounted remote control stations (KS RCS/1 and /2) are die-cast aluminum with a stainless steel front plate, tempered against opening, and with a red and green LED (Fig. 19-77). The waterproof version ensures unit use during all types of weather conditions. Figure 19-78 provides basic wiring information for the RCS/1 and /2 units.

Another remote control station with an anti-tamper switch is at Fig. 19-79. Like the other units, it uses the Mini-Kaba or the Kaba 20 key/switch. A remote control station mounting box (Fig. 19-80) is available and used for mounting the unit. Mounting box dimensions are at Fig. 19-81.

Remote control switches are UL approved. Electrical specifications are at Fig. 19-82.

block" on the programmer wiring hookup. Use this only for duress/error entry.

To program the Memorilok, the programmer key is turned to the on position. Enter the desired four digit code combination on the keyboard. A green LED indicator light will glow to indicate that the code is properly programmed into the memory system. Return the programmer key to the off position. Reinsert the code to test the unit for proper operation. Now the green LED light will indicate acceptance of the code combination and supply of power to the lock or other device used.

The adjustable time delay for the Memorilok affects the lock open time (time delay for the unit) and it can be varied from one to ten seconds by turning an adjusting screw.

Unit specifications are at Fig. 19-72. Specification installation details, template, and other specific information concerning conductor wire are provided with the unit from the factory.

Chapter 20

The Business and Law of Locksmithing

Locksmithing is a profession involving skill, competence, and public trust. It is a career that can be highly rewarding. This chapter will give you the background necessary to use the basics of what you have already learned and apply it in a business.

The *Occupational Outlook Handbook* indicates that in 1972 there were approximately 8000 full-time locksmiths in the United States. Many of these 8000 individuals operated their own locksmithing business, but others were in a shop which employed one to three persons. The situation is very much the same today, meaning that in this country, the demand for competent locksmiths is overwhelming.

Fifty to 75 million keys are lost and must be replaced each year. Every household, automobile, and business has two to five locks. In a medium-sized city of 75,000 to 100,000 persons, there are easily over 250,000 locks. This means there are literally billions of locks and keys in this country—plenty of business for a good locksmith. Statistics show that locksmiths make good money, too.

BUSINESS CONSIDERATIONS

If you're thinking about launching a career in locksmithing, there are several things you should consider:

☐ *Police clearance*—Many jurisdictions require licensing and fingerprinting.

☐ *Professional qualifications*—Some cities insist on tests to determine professional competence.

☐ *Basic tools and equipment*—You should insure that you have all you need to tackle any locksmithing job.

☐ *Supply stock*—It's possible to work out special arrangements with some local supply houses.

☐ *Financing*—Setting up shop involves an initial cash outlay.

☐ *Bookkeeping*—Records must be kept of all transactions.

☐ *Vehicles*—Necessary for out-of-shop calls.

☐ *Advertising*—Sometimes it takes more than just locksmithing competence to bring in the business. Good advertising usually includes signs, business cards, window displays, display racks, display materials, etc.

☐ *Literature*—Locksmith supply house catalogs, manufacturers' literature, reference books, business forms, etc.

☐ *Contacts*—It pays to know other locksmiths, factory representatives, etc. They can help you from time to time with locksmithing problems.

SELECTION OF THE BUSINESS SITE

Selecting the business site is a major decision, and one that will affect the business throughout its infancy.

In determining the location, consider the types of businesses you will be near and what effect they will have upon your customers. Shopping malls are ideal for locksmithing shops because of the walk-in business generated by people who need one or two keys duplicated. Should you undertake home servicing, you would want a location that enables you to get to your customers quickly.

The type of building must also be considered. It would be unwise to obtain a building that is in need of massive repairs. The building should also be large enough for your needs. Thought should be given to the building's potential for future expansion.

Demographic and economic factors of the surrounding area should be considered too. Make a study of population density, population growth, projected economic growth, the tax structure, etc. Examine any potential competition with other locksmiths.

SHOP LAYOUT

Window displays are a good way to attract customers. The display itself should be simple and effective, neat, and clean. Window displays cannot be cluttered; you must be discriminating and orderly in developing them. The display should tell the viewer something— something interesting and appealing about your products or services.

The front counter area should be carefully organized. This is where sales are made. Ensure that the key machines are near the counter—visible from the street. Spread out your keys on a key board; such a display stays in your customers' minds longer than a pile of keys on a table. Code books usually are kept near the key machines.

The workbench usually should be out of the customers' sight, preferably in the back room. If it is out front, it should be clean at all times.

Lighting of the work and customer service areas is most important. It should be adequate but not over-lighted. The lights should be of the nonglare variety. The work area can have additional small lights placed several feet above the bench.

MERCHANDISING THROUGH ADVERTISING

Advertising takes many forms, from radio, television, and handbills to putting your name on each key duplicated. There are no hard and fast rules. Good advertising is advertising that gets results.

The selection of the proper advertising can be important for a beginning locksmith. Consider the population; do they listen to the radio, watch television, or read the paper more than in other areas, or do they look for gimmicks? What are the costs of the various advertising media? You can approach the advertising staffs of the local media to determine these answers.

Once an advertising scheme is carried out, evaluate the results. Did the advertising increase business? Was the advertising worth the investment?

Ultimately, the locksmith himself is the most effec-

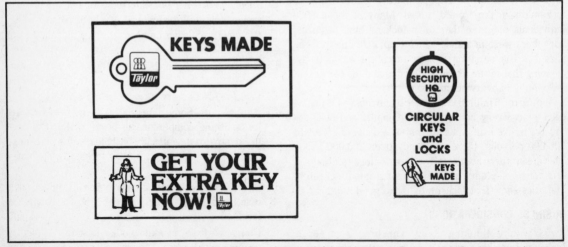

Fig. 20-1. Various sales aids for use in the locksmith shop. At upper left is a key sign; others are window banners. (courtesy Dominion Lock Co.)

Fig. 20-2. Advertising sign to hang in window of locksmithing shop. (courtesy Dominion Lock Co.)

tive advertising. Responsible locksmithing, a neat appearance, and a little courtesy go a long way. In spite of all the sales talk and advertising hoopla, the customer is only interested in one thing—quality service.

Whether your work as a locksmith is out of your home, your vehicle or out of a formal locksmithing shop, you need to have some form of 'advertising' at your shop. To have a sign outside that merely states "Locksmith" isn't really enough. Whenever possible, you should have and use the small stickers that can be affixed to windows, the side of a van truck, or whatever, as they can increase your business.

Figure 20-1 shows a variety of such aids, including decals, window banners, and a sign. Whenever possible, you should be using these types of aids. They promote business and tell your potential customers exactly what services you can or cannot provide. Such signs also save you time in answering general inquiries from walk-ins off the street who are not sure if you can provide a specific service, such as cutting a circular ("Ace" type) key. In this instance, having a sign that indicates that these keys are duplicated in your shop can increase business.

A hanging sign (Fig. 20-2) with a slightly different slant to it—"Keys Cut While you Watch"—is another visual aid to customers. Again, by using the properly aligned key machines, you can ensure that the phrase "guaranteed fit" is accurate.

Decals for your office or on the service van also aid in promotion of your business and security in general (Fig. 20-3).

KEY BLANKS AND THEIR SELECTION

As a locksmith, it is important to carry a variety of key blanks to meet the continuing needs of your customers. More importantly, you must be able to determine what *blanks* are most used and popular in your area. For you to purchase one or two boxes of every type of key blanks imaginable that *might* possibly be used during the course of a career in locksmithing is ridiculous and extremely costly. Purchasing such items willy-nilly based upon just any key catalog that comes across your desk shows poor business judgment. The large companies that supply all kinds of locks and security devices will, naturally, also have the widest possible variety of key blanks available for your purchase and use. That assortment doesn't mean that you really need such an extensive variety of key blanks. In almost every case, you do *not*!

To consider what you *should* have, let's view the key determination process from the viewpoint of a key blank manufacturer. The following information comes from Herbert Stein, president of Star Key & Lock Manufacturing, Co., Inc., of New York. This company has been in the business of carefully foreseeing the key blank needs of locksmiths for quite a number of years.

Fig. 20-3. Decals for use in office or on side of van. (courtesy Dominion Lock Co.)

The first step is to consult a number of the largest competitors in different parts of the country for their opinion of the relative popularity of a particular blank and the anticipated annual demand. (Some blanks may be extremely popular in certain areas of the country, but very slow in other parts.) On the basis of this private poll, a decision will be made. This decision is also influenced considerably by such factors as whether the blank's lock is made by a major company or not, and whether it is a big volume lock or a small one (for example, if the blank is for an automobile, is it for the major line of one of the top domestic or foreign automobile makers, or is it just for one of the small volume imported cars?). Naturally, if a deluge of requests come in from across the nation requesting a specific type blank, the priority for putting that one into immediate production will go to the top of the list.

Now, knowing how a key blank is selected for production helps, but how do you go about finding out what the most common blanks are in your area? Here is where your regional representative of the manufacturer, or direct contact with the factory, can help you. Knowing what keys are most used in your area is part of their job. Through various national and regional trends, regional and area market studies, and previous sales to your area, the manufacturer can determine with great accuracy what the most common and popularly requested blanks will be for any given part of the country.

While manufacturers keep such records, you may feel that you are imposing on them with a "trifling request." Not so, but even then, some locksmiths just look through a catalog, especially when starting out in a business, and contact a local locksmith distributor/supply house in the area. The supply house can assist you, but many sales of key blanks don't go through the supply house; they are purchases from other area distributors or directly from the manufacturer. So the supply house, while in touch with your area, isn't always as fully cognizant as the manufacturer in this regard.

At the supply house you will probably be shown one or several catalogs with a myriad of key blanks with various ones indicated that you should have on your shelf. These may or may not meet your requirements. Again, Star Key & Lock has thought of this; their catalog lists the most commonly used key blanks for just about every situation. This is where dealing directly with the manufacturer is to your advantage. By purchasing assortments of different types, you reduce your initial key blank costs, reduce your stockroom inventory, have more space for other products, and know exactly what your customers will expect. In addition, you don't have several hundred or more blanks to look through for any given type, but perhaps only a hundred and fifty or so.

By purchasing in an assortment mode, you have the advantage of getting the keys in display mounts that also provide a subtle buying hint to the customers who walk into your shop. These displays hold the *most* popular key blanks (usually between 90 and 150 different) that will meet just about all of your needs. The few oddball keys that come in for duplication are also the ones that give every locksmith a headache, because it may be only once every two to five years that a key of that specific type is ever needed.

A little footwork at your end helps you assess the needs of your shop. Check out the local general supply houses and see what blanks are the most popular with them. Also ask them about the most commonly purchased locksets in your area, and get to know these locks, the different keyways possible and available for them, and also what *were* the big sellers within the past ten years— not just the big sellers in today's market.

From all of this, you should have learned:

☐ The most popular key blanks in your area.
☐ The volume of blanks used in the past several years for different types of locks.
☐ The top dozen locks, by manufacturer, selling currently in your area.
☐ The top dozen locks, again by manufacturer, that were big sellers during the past decade.
☐ The various keyways necessary for these locksets.

This knowledge (coupled with the size of your shop and projected business) gives you a very good idea of which key blanks will be the most in demand, which ones will require a medium stock, and which ones you have to consider only once every two or three years. The minimal stock blanks will last for a long time, but the more popular ones may need frequent restocking—possibly every month.

With all this research done, you can now properly select the blanks that you feel are necessary. Remember that it is probably best to consider the general assortment of key blanks when first starting out. Get the catalogs that provide these types of blanks and consider their assortments. Later on, as your business expands, you can move on to include other key blank manufacturers for other, less common blanks that you see a need to stock.

After completing your key blank research, you should now know the types of locks required, as well as exactly what type of key reference catalogs for key codes your shop should carry. You can also start listing the specific cylinders, pin sizes, and dimensions required, specific repair parts kits, key machines that may be necessary (say, for cutting Ace keys), and some other specifics that will be necessary to perform as a locksmith.

The professional locksmith knows his (or her) key blanks and doesn't need to constantly refer to charts and catalogs for this information. In the beginning, every locksmith has trouble identifying one key among hundreds that may be on the shop wall. Star Key & Lock has come up with a nifty idea that is extremely helpful (and something that I wish that all manufacturers of key blanks would do): they put an "instant identity on the key blank. This new feature has put Star far out in front of other manufacturers. What it amounts to is "stamping the make of the lock" on the key blanks in addition to the comparative numbers. This achieves immediate identification of the key blank by the locksmith, saving time and trouble, and is a valuable learning tool for the student and apprentice locksmith. The make of lock stamped directly on the key also helps convince the customer that he is receiving the correct key for his lock, which aids in establishing the locksmith's credibility.

An added feature that you as a locksmith can supply to customers who require something special or an identifying logo is embossing. This means imprinting the bow of the key blank with anything from a fancy logo design to several lines of advertising copy. Imprinting can be done on any number of the brass keys available from Star. I mention this because it is not a common or standard service, though some other manufacturers offer it as well.

Special embossing usually consists of three lines or less of standard type copy. This is based on a maximum of 14 figures on the first line, 12 on the second line, and 9 on the third line. The manufacturer must make back the costs of embossing; as a minimum, consider an order of 1000 embossed keys of assorted key blank numbers in increments of at least 50 pieces per number ordered.

The embossing is an excellent way to advertise your business. It costs less than average advertising, and still gets your message across to the public that needs and uses your products and services (Fig. 20-4). The chief advantages of special embossing are excellent advertising, increased locksmith shop prestige, and indisputable proof that the keys were made by the name embossed thereon.

In closing out this discussion of key blanks, I must comment that the keys in use today, for the most part, are of brass (all brass); this is important. All-brass key blanks

are best for impressing because they have structure which will show distinct impression marks. The blanks cut very easily on key machines or by hand filing, but the myth that the brass is soft because of this is not true. The all-brass key retains maximum strength (77,000 p.s.i.); the special brass alloy which is used by Star Key & Lock (and other manufacturers) gives the best combination of desirable properties of impression-sensitive structure, free-cutting duplication, and the maximum strength necessary for a key blank.

Both as a locksmith and as an individual who requires keys in your own day-to-day movement (home, auto, office), you will see through experience that certain key blanks are better than others. Use the products that are of the best quality, not those that cost the least!

A key machine for the office or for use on the job in remote locations is shown at Fig. 20-5. This compact, highly dependable cylinder key cutter, the "Companion," will go wherever your locksmithing work may be. It has dual voltage (110 Vac or 12 Vdc capability with no extra wiring necessary; for traveling situations it comes in a sturdy carrying case and weighs only 25 lbs.

The machine itself has four-way jaws to grip any cylinder key; no special adapters are required. The precision alignment of the cutter ensures a correctly duplicated key everytime. (Of course, you should always check the cutter alignment periodically to ensure that the precision alignment is true. Rough handling and occasional bumps may throw the alignment out slightly. Be safe, be sure, and cut a true duplicate key every time.)

Fig. 20-4. Excellent advertising increases the locksmith shop prestige and also provides indisputable proof that the keys were made by you. (courtesy Star Key & Lock Co.)

Fig. 20-5. The "Companion" model portable key machine operates off 110 Vac or 12 Vdc current. (courtesy Dominion Lock Co.)

The key display and machine is a visual selling aid in the locksmith shop. Usually located near the counter or cash register, it promotes sales subtly by showing your customers that you have the commonly used keys in stock and can quickly duplicate a key for them while they wait. In the case of the display at Fig. 20-6, it is important that you know exactly what is available. This particular key assortment and cutting machine is a self-standing revolving counter rack and covers both U.S. and foreign automotive keys in addition to keys for business and residential locks.

The jaws on this key machine can be rotated to present four different gripping surfaces to cut four different types of keys without using any adapters. The jaws simply lift up and rotate, eliminating the fumbling necessary with adapters for the key machine. The "standard" station of the cutter jaws holds regular cylinder keys, while the "wide" jaws grip Ford double-sided keys. The "A" station jaws hold the Schlage wafer key (SC 6) and the "W" station jaws are for the Schlage wafer key (SC22). Other keys, such as Chicago and double-sided Datsun keys, can be cut with the "A" station jaws.

KEY MACHINE CUTTERS

After continued use, the cutters for your key machine(s) will become dull. This, in turn, means that

Fig. 20-6. Key machine with the most commonly cut key blanks available for immediate duplication for customers in need. (courtesy Taylor Lock Company)

Fig. 20-7. A set of the basic replacement key machine cutters: file, slotting, milling, and a side-milling slotter. (courtesy Taylor Lock Company)

each key you cut will not have the precision cuts required. Any such key may operate roughly in the lock, or not at all. To alleviate this, you should have at least one replacement cutter on hand of each type you normally use.

The basic cutters are shown at Fig. 20-7, and include the file, slotting, milling and the side milling slotter. The cutter specifications and key type uses are as follows:

Material	Used For	Cutter Type
File	High carbon steel	Cylinder keys
Slotting	High speed steel	Flat keys with .045″ cut*
Milling	Tungsten-chrome alloy steel	Cylinder keys
Side Milling Slotter	High speed steel	Cylinder and flat keys (.045″ cut)

*Two different slotting cutters are available with different thicknesses for .030″ and .045″ cuts on flat steel keys.

DISPLAYS

Shown in Fig. 20-8 is the color key carousel. Because colored keys are becoming more popular, some customers will prefer to have another color key when having their key duplicated. Color keys assist the customer in rapid identification of specific keys. In some instances, they will not accept a key that is not colored.

A key comparison board allows you to rapidly determine which blanks, by manufacturer, are the same in their component makeup (i.e., the specifics of the key identification). A key comparison board is a great advantage to the locksmith, and even more so for the apprentice or locksmith trainee.

In the case of these two items (as with others that you will find in various catalogs), the maker isn't just hyping a product in hopes that you will buy it; the man-

Fig. 20-8. Color key carousel from manufacturer to assist you in the locksmithing business. (courtesy Star Key & Lock Co.)

Fig. 20-9. Manufacturer-prepared lock display for customer consideration. Every working shop should have a variety of such displays for the customer. (courtesy M.A.G. Engineering and Mfg., Inc.)

ufacturer genuinely wants to provide you, as a locksmith, with items that will assist you in the locksmithing trade.

By helping you to do your job a little better, easier, faster, and more professionally, and by making available the supplementary business items required, the manufacturer is helping everyone: you, your customer, and himself. When you have problems, questions, or need assistance, the manufacturer or his representative is there to help you. Likewise, the manufacturer is looking ahead to your varying needs and has a variety of items to help you to help yourself.

Another example of a manufacturer's concern for assistance to the locksmithing business is at Fig. 20-9. Lock displays like this unit are available for setup and demonstration of specific products within your shop. The customer is interested in a specific type of product—in this case, a lock. With a variety of these type of displays available, the customer can try out each one, and as the choice of locks is narrowed down, you provide specifics concerning the remaining locks. In this particular case you might point out the very salient and positive features of the Ultra 800, the Medeco cylinder, the strength and positive locking action of the mechanism, and other specifics.

For narrow stile door installation, Adams Rite has created a specific installation kit. The advantages of this kit should be obvious: extra parts and detailed information for each stile installed. For the locksmith who plans continued installation, repair, and/or replacement of lock units in narrow stile doors, this kit is a must. It allows for a smoother, more rapid installation of numerous units, saving a tremendous amount of time (Fig. 20-10).

THE LOCKSMITH AND THE LAW

The locksmith, because of the uniqueness of his profession, must have a better understanding of various laws than most people. In this regard, when setting up a locksmithing business it is always prudent to consult with your attorney regarding all laws that concern you as a locksmith. In many jurisdictions there are laws covering licensing, control of locksmithing tools, and registration of code books. Some local laws regulate the conduct of certain locksmithing business practices, such as duplicating master keys, making bank deposit box keys, opening automobiles, etc.

When first entering the business, aside from consulting with your lawyer and possibly the police, you should also contact your area locksmithing association, if one exists. Information can be obtained from the national locksmithing organizations too, through their newsletters and publications. These are excellent sources of information about your legal responsibilities.

Your legal responsibilities demand that you be very careful about the jobs you accept. You must be certain that you are not breaking the law by complying with a customer's wishes. Authorization statements from supervisors, such as from a bank or post office, for duplicating a key should always be doublechecked with the main offices. Verification of such written statements is a must. They should also be filed with the job order.

When jobs of this type arise, the following information should be included in your files:

- ☐ Name of person who brought the job in.
- ☐ Identification (Social Security card or driver's license).
- ☐ Address of person who brought the job in.
- ☐ Business telephone (call to doublecheck).
- ☐ Name of firm.
- ☐ Business address.
- ☐ Type of service performed.
- ☐ Type of payment (if a check, it should be a business check, not a personal check).
- ☐ Automobile make, model, serial number, license number, state it is registered in.

You may also have some forms printed up for the individual to fill out. A single form can be used to cover all the

330

Fig. 20-10. Installation kit for the Adams Rite narrow stile door units, a must item involved in this type of lock installation. (courtesy Adams Rite Mfg. Co.)

situations you may run into. The form should be kept along with the work order.

In some states or cities, laws require that you take an examination before your locksmithing peers to show that you are a competent locksmith meeting professional standards. Besides this, police checks of your reputation, qualifications, background, and previous employers may be made to ascertain that you are of good moral character.

Various laws have been enacted to protect the public from unscrupulous individuals posing as locksmiths and to protect the locksmithing trade itself from such individuals. The following is extracted from the Los Angeles Code, Ordinances No. 83,128, as an example of a local law regulating the locksmithing profession:

SEC 27.11 LOCKSMITH—REGULATIONS APPLICABLE TO

A. Definitions

"Locksmith" shall mean any person whose trade or occupation, in whole or in part, is the making or fashioning of keys for locks, or similar devices, or who constructs, reconstructs or repairs or adjusts locks, or who opens or closes locks for others by mechanical means other than with the regular keys furnished for the purpose by the manufacturers of the locks.

B. Trade of Locksmith—Permit Required

No person shall engage in the business of locksmith, or practice or follow the trade or occupation of locksmith without a permit therefore from the Board of Police Commissioners.

C. Permit—Application

Such permit shall be issued only upon the verified application of the individual seeking the permit. The application shall be upon a form prescribed by the Board and shall set forth the proposed location of the applicant's place of business, the names and addresses of five character references, and such other things as the Board may require to determine the character, honesty, and trustworthiness of the applicant. Specimen fingerprints of the applicant shall be furnished with the application.

D. Permits—Fees—Expiration

Each application for a permit shall be accompanied by a fee of $10 and each application for the annual renewal thereof, by a fee of $5. Each permit, unless sooner revoked, shall expire on December 31st, following the date of issuance. Each permit shall bear a serial number.

E. Permits—Issuance and Denial

The Board shall cause an investigation to be made upon each application and if the Board finds that the applicant's reputation for honesty is good, that he has not used his skill or knowledge as a locksmith to commit or aid in the commission of burglaries, larcenies, thefts, or other crimes, that he intends honestly and fairly to practice the trade of locksmith in a lawful manner, and that he has not been convicted of a felony, then the permit shall issue. Otherwise it shall be denied.

F. Permittee—Must Keep Record

Each permittee must keep a book, which shall be open to inspection by any police officer at all times, in which the following must be entered:

1. The name and address of every person for whom a key is made by code or number.
2. The name and address of every person for whom a locked automobile, building, structure, house, or store, whether vacant or occupied, is opened, or a key fitted thereto.

G. Keys to Be Stamped

It shall be unlawful for any locksmith to fail to stamp the serial number of his permit upon any key made, repaired, sold, or given away by him.

H. Signs to be Displayed

Every locksmith shall display in a conspicuous manner in the place where he is carrying on such business, trade, or occupation, a sign of a style, size, and color to be prescribed by said Board, reading, "Licensed Locksmith," together with the official permit number.

I. Permits—Revocation

The Board, upon proceedings had as prescribed in Section 22.02, may revoke or suspend any permit issued hereunder upon any of the following grounds:

1. Misrepresentation in obtaining such permit.
2. Violation of any provision of this section.
3. That the permittee has committed or aided in the commission of or in the preparation for the commission of any crime by the use of his skill or knowledge as a locksmith or by using or letting the use of his tools, equipment, facilities, or supplies.

Selecting a Padlock

Though initially developed by the Master Lock Company with their products in mind, this appendix is a real gold mine of information concerning what anyone should look for when considering the purchase of any padlock.

What is the real key to security when you pick out a padlock? The fact is that locks vary tremendously in the amount of protection they provide. Important differences may be hard for you to spot. Keep the following in mind when you're shopping:

Wrought-Steel Padlocks—You Get What You Pay For

Wrought-steel or "shell" padlocks are the lowest-priced locks you can buy. A great buy if you understand they're intended mainly for nuisance protection—keeping the kids out of your tool box, locking power tools against tampering, restricting access to a mailbox, storage room, cage, etc. Costing little, they can prevent thoughtless misadventures or injury from hazardous household and industrial items.

Warded Padlocks—Good Strong Middle Ground

Laminated padlocks are far sturdier than shell types—much more resistant if anyone tries to hammer

open the case. Design permits a wider range of key variations than offered by shell locks.

Costing more than shell types—but less than those with pin-tumbler mechanisms—warded padlocks give dependable low-cost pilfer protection. Use them where property is of significant but limited value—securing oil tank caps, well covers, beach lockers, duffel bags, barn doors, etc.

Because of the relatively large clearances between internal moving parts, warded padlocks are frequently chosen for applications where sand, water, ice, or other contaminants are a problem.

Where corrosion eats at padlocks, brass has better staying power. Hard-wrought brass padlocks are designed to withstand severe corrosion problems encountered along seacoasts, aboard boats, at refineries, and in areas of high humidity and atmospheric pollution. For theft deterrence, the shackle will typically be chrome-plated hardened steel. But for extreme conditions, even the shackle can be brass. Expect the price of maximum-security brass padlocks to run substantially higher than for steel.

The solid-brass padlock is a lower cost medium-security alternative for the average user. Again, look for pin-tumbler locking and case-hardened shackle to protect valued property. Priced substantially less than heavy-duty laminated brass padlocks, solid-brass locks are naturals for many coastal uses, boats, outdoor lockers, gates, and similar applications.

Combination Padlocks—the Key Is in Your Head

The chief reason for this kind of padlock is keyless convenience—particularly with children where lost keys may be a problem. Protection features to look for include reinforced double-wall construction—tough stainless steel outer case over a sturdy wrought-steel innercase—and hardened steel shackle for added resistance against cutting and sawing. Combination padlocks are classed as medium security, with strength equivalent to a quality warded padlock.

No Stronger Than the Weakest Link

People should consider more than just the strength of the padlock they choose. The finest lock affords little protection if burglars find it hung from an undersized or unhardened hasp that can be cut with ease. To avoid a weak link in security, get a hasp that matches the quality of the lock. Look for adequate size, pinless hinge, concealed screws, case-hardened staple (the metal loop the lock passes through), and steel ribbing for added strength.

An evolutionary step beyond the hasp is the hasplock. Here instead of using a separate padlock, the lock and hasp are one, permanently joined so the lock can't be misplaced or stolen. Hasplocks give built-in convenience akin to the deadbolt door lock while offering much easier installation. This accounts for their popularity with builders, contractors, farmers, and do-it-yourselfers. Their uses range from garage doors and sheds, to boat houses, truck warehouses, and a host of other places.

Another weak link in security may be the wrong choice of chain or cable. When it comes to movable property, the use of a strong padlock with a properly matched chain or steel cable is the way to go. They enable owners of bicycles, boats, motorcycles, and other easily stolen goods to obtain maximum protection by securing them to a post, tree, dock, or other immovable object. Common chains available in hardware stores should be passed over in favor of chains specifically designed for locking applications. Be sure its case hardened for high resistance to cutters, saws, and files. Individual links should be welded, not just twisted, to resist being pried apart. Multistranded security cable is available for equivalent protection, with the added benefit of light weight. In either case, the thicker the chain or cable, the greater the protection. Examine cable closely: Some manufacturers add a thick coating of vinyl to make a small steel strand look bigger.

For greatest protection, position the lock and cable (or chain) as high above the ground as possible. This makes it difficult for thieves to gain extra leverage by bracing one leg of a bolt cutter against the ground.

When in Doubt About a Lock...

The best place to turn with questions of security is your local police department. Most law enforcement agencies have experts to help you and may have displays of security locks. When in doubt and unable to get expert advice, the rules to know are:

1. Buy the best padlock protection you can afford (strength and cost generally coincide).

2. When possible, avoid leaving locked items in out-of-the-way places where thieves have time to work on the lock unseen.

3. If locked property has strong appeal to the light-fingered, back up your locks with some insurance.

Locksmithing
Manufacturers and Suppliers

When requesting specific production information, prices, or just a current catalog from a manufacturer or supplier or locksmithing products, please use your locksmith letterhead stationary, and mention this book as a basic source of reference. When ordering, check first with your area or regional distribution supply point, who may be a direct factory representative or locksmithing supply house.

Where indicated by an asterisk (*), contact with the factory may be made directly, especially in an area or region where there is no local distribution point for the product.

*Adams Rite Manufacturing Co.
P.O. Box 1301
4040 South Capital Avenue
City of Industry, CA 91749

*Alarm Lock Company
5411 Telegraph Road
Los Angeles, CA 90040

Allen and Company
10 Smithfield St.
Pittsburgh, PA 15222

*Commonwealth Lock Company
1853 Massachusetts Avenue
Cambridge, MA 02140

Dominion Lock
Air Industrial Park
Plattsburgh, NY 12901

*Kemko Supply Company
Box 4151, Benson Station
Omaha, NB 68104

*Lawrence Locksmith Supply
300 Long Beach Road
Oceanside, NY 11572

*Lock Corporation of America
6301 West Mill Road
Milwaukee, WI 53218

*M.A.G. Engineering & Mfg., Inc.
13711 Alma Avenue
Gardena, CA 90249

M. K. Morse Co. (The)
P.O. Box 8677
Canton, OH 44711

*Preso-Matic Lock., Inc.
3048 Industrial 33rd St.
Fort Pierce, FL 33450

Security Engineering, Inc.
61 E. Main St.
P.O. Box 337
Forestville, CT 06010

*Sentry Door Lock Guards, Inc.
P.O. Box 326
Dania, FL 33002

Simplex Security Systems, Inc.
Front and Main Sts.
Collinsville, CT 06022

Star Key & Lock Mfg., Co., Inc.
1274 Flushing Avenue
Brooklyn, NY 11237

Taylor Lock Company
2034 Lippincott St.
Philadelphia, PA 19132

Trine Consumer Products Div.,
Square "D" Company
1430 Ferris Place
Bronx, NY 10461

Plug Follower and Holder Diameters for Popular Locks

Manufacturer	Diameter (in inches)			
	0.395	0.495	0.500	0.550
Acrolock		X		
Corbin		X		
Corbin oversize			X	
Corbin small pin	X			
Eagle		X		
Eagle small pin	X			
ILCO		X		
ILCO 4019 rim cylinder			X	
ILCO "peanut" cylinder	X			
Keil			X	
Kwikset		X		
Lockwood		X		
National (Rockford)		X		
National (Ozone Park)		X		
Russwin oversize			X	
Sargent		X		
Segal		X		
Taylor		X		
Weslock (except KNK)		X		
Yale		X		
Yale small pin	X			

Glossary

Ace lock—A lock in which the tumbler pins are arranged in a circle.

antipick latch—A spring latch fitted with a parallel bar that prevents the latch from responding to external pressure from a tool that is not a proper key.

armored front—A plate covering the bolts or set screws holding a cylinder to its lock. These bolts are normally accessible when the door is ajar.

backplate (rim cylinder)—A small plate applied to the inside of a door through which the cylinder connecting screws and bar are passed.

backset (of a lock)—The horizontal distance from the face of a lock to the center line of its cylinder, keyhole, or knob hub. On locks with beveled fronts, this distance is measured from the center of the face. On locks with rabbeted fronts, it is measured from the upper step at the center of the lockface.

barrel key—A key with a round stem that is hollow at its end. The hollow fits over a pin in a lock keyway and helps keep the key aligned. Also known as a hollow post key or pipe key.

bicentric cylinder—A special lock having two pin-tumbler cylinders. The correct key in either cylinder opens the lock. This permits master keying without reducing the security of the locking mechanism.

bit—That part of a key that is cut to fit a lock.

bit key lock—A lock operated by a key having a wing bit.

bitting—A cut, or series of cuts, in a bit.

blank—A key that has not been cut or shaped to fit a specific locking mechanism.

bow—That portion of a key that is held between the fingers when the key is being used.

burglarproof—A term used to describe locks or doors designed to be impregnable against a thief without explosives or an unlimited amount of time.

burglar resistant—A term used to describe locks or doors capable of resisting attack by prowlers and thieves for a limited time.

cam—That portion of a lock that turns when its key is turned or when the bolt knob is turned. Unlike a tang or tail, a cam has no play and operates around a central axis.

case (of a lock)—The box containing the lock-operating mechanism.

case ward—A ward or obstruction integral to the case of a warded lock.

change key—A key that operates only one lock in a series, as distinguished from a master key which operates all the locks in a series.

clean opening—A locksmith's term for opening a lock without obvious force, i.e., a skilled entry.

closet spindle—A knob spindle having a turn knob made fast to one end and with provision for attaching a knob to the other end. For use on the inside of closet doors.

combination lock—A lock that does not use a key but requires that certain parts be placed in a specific juxtaposition by use of a turning knob on the face of the lock. Generally the parts are coded with digits and the correct digit series sets the internal parts in the right order so the lock will open.

connecting bar—A bar attached to the cylinder rear of a plug lock to operate the locking bar mechanism.

core—The plug of a cylinder lock.

corrugated key—A key with pressed longitudinal corrugations in its shank to correspond to an irregularly shaped keyway.

cuts—The indentations made in a key blank in order to fit the key to a lock.

cylinder—A lock housing containing a tumbler mechanism and a keyway.

cylinder lock—A lock having mechanisms operated by a cylinder.

cylinder ring—A collar or washer used under the head of a cylinder.

cylinder screw—The set screw in the front of a cylinder lock which prevents the cylinder from being turned after installation.

deadbolt—A lockbolt having no spring action, usually rectangular and actuated by a key or turn knob.

deadlatch—A lock with a beveled latchbolt that can be automatically or manually locked against end pressure when projected.

deadlock—A lock with a deadbolt only.

depth key—A special key that enables a locksmith to cut blanks made for a special lock according to a code.

derivative code—A special numerical code that relates a lock's tumbler arrangement to the depth of the key cuts necessary to operate the lock.

disc tumbler—A circular or oval-shaped disc with a rectangular hole and one or more side projections. A number of these are used side by side in a disc-tumbler lock.

door check—A device consisting of a heavy spring and arm coupled to an air or oil cylinder that prevents the spring from closing the door too rapidly. Also called a door closer.

double-bitted key—A key with cuts on both sides of its blade.

drill pin—A round pin projecting from the inside of a lock case to receive a barrel key. Also called a *barrel post*.

drivers—The upper set of spring-activated pins in a pin-tumbler cylinder lock.

dummy cylinder—A cylinder containing one operating mechanism.

emergency key—A key capable of opening any lock in a building even though a door may be locked from the inside.

escutcheon—A plate, either protective or ornamental, containing openings for all of the controlling members of a lock such as knob, handle, cylinder, keyhole, etc.

fence—Another name for the post in a lever-tumbler lock.

finished key—A key that has been cut to fit a lock.

following tool—A tool used to hold the pin tumblers and springs in place while a pin-tumbler cylinder is being assembled, disassembled, inserted, or removed.

gate—The opening in lever tumblers that allows them to pass the post, or fence. Also called gating.

graphite—A lubricant especially useful for the finer parts of a lock.

heel (of a padlock shackle)—The end of a padlock shackle that is not removable from the case.

jamb—The inside vertical face of a doorway or window frame.

keeper—Another term for strike.

key blank—An uncut key. Once cut, a key blank becomes a finished key.

key caliper—Any small caliper capable of measuring the height of a cut on a key.

key extractor—A device used to remove broken key pieces that have broken off in a lock keyway.

key plate—A small plate or escutcheon having a keyhole only.

keyway—The opening in a lock mechanism into which the key is inserted.

latch—A door fastening device, usually with a sliding or spring bolt.

latchbolt—A beveled spring bolt, usually operated by knob, lever handle, or thumb piece.

layout board—A board with a number of parallel grooves used to hold pin-tumbler parts in order when the locksmith is working on the lock.

lever handle—A horizontal handle for operating the latchbolt of a lock.

lockface—A plate that shows in the edge of a door after lock installation.

locking dog (of a padlock)—The part that engages the shackle and holds it in the locked position.

master key—A key capable of operating several different locks, each lock being operated by its own change key.

mortise—An opening made in a door to receive a lock or other hardware. Also the act of making such an opening.

mushroom pins—A mushroom-shaped driver.

night latch—An auxiliary lock with a spring latchbolt and functioning independently of, and providing additional security to, the regular lock on the door.

panic bolt—A type of lock fitted with a long bar placed horizontally across the inside of a door. Pressure on the bar releases the lock. Used in theaters, schools, and other public buildings. Also called panic bar.

pass key—A master key or skeleton key.

pick—Any tool or device (other than a key), either from a locksmith supply house or homemade, that is used to open a lock.

pin tumblers—Small sliding pins in a lock cylinder that work against coil springs and prevent the cylinder plug from rotating until the correct key is inserted.

plug (of a lock cylinder)—The cylindrical mechanism in a lock cylinder that houses the keyway.

plug retainer—A device that retains the plug in a cylinder lock. Also called a retaining ring.

post (of a key)—The cylindrical portion of a bit key to which the bit is attached.

reversible lock—A lock in which the latchbolt can be turned over so as to adapt it to doors of either hand, opening in or out. This does not apply to beveled front locks and certain cylinder mortise locks.

security—In the locksmithing trade, the ability of a lock to withstand attempts at unlawful entry.

Shackle—The curved portion of a padlock that passes through the hasp.

shackle spring—The spring that projects the shackle from a padlock case when unlocked.

shank (of a key)—The part of a bit key between the bow and the bit.

shear line—The space between the shell and the plug of a lock cylinder.

skeleton key—A warded lock key cut especially thin to bypass the wards in several warded locks so the locks can be opened.

spacing—The distance between a lock's keyhole and its spindle hole.

top pin (of a padlock)—A pin that retains the shackle in a padlock case when unlocked.

trike—The part of a locking arrangement that receives the bolt, latch, or fastener. usually recessed in the door frame. Sometimes called a keeper.

tension wrench—A device used to maintain constant pressure on a lock while its tumblers are being manipulated by a pick. It assists in picking the lock and moving the plug in the lock after it has been picked.

thimble—A tool used to hold the plug of a pin-tumbler lock while it is being worked on.

Index

Other Bestsellers From TAB